Organic Chemistry at a Glance

Laurence M. Harwood
Professor of Organic Chemistry, School of Chemistry,
University of Reading

John E. McKendrick
Lecturer in Organic Chemistry, School of Chemistry,
University of Reading

Roger C. Whitehead
Lecturer in Organic Chemistry, Department of Chemistry,
University of Manchester

Blackwell
Publishing

© 2004 by Blackwell Publishing Ltd

Editorial offices:
Blackwell Publishing Ltd, 9600 Garsington Road, Oxford OX4 2DQ, UK
 Tel: +44 (0)1865 776868
Blackwell Publishing Inc., 350 Main Street, Malden, MA 02148-5020, USA
 Tel: +1 781 388 8250
Blackwell Publishing Asia Pty Ltd, 550 Swanston Street, Carlton, Victoria 3053, Australia
 Tel: +61 (0)3 8359 1011

The right of the Author to be identified as the Author of this Work has been asserted in accordance with the
Copyright, Designs and Patents Act 1988.

All rights reserved. No part of this publication may be reproduced, stored in a retrieval system, or transmitted, in
any form or by any means, electronic, mechanical, photocopying, recording or otherwise, except as permitted by the
UK Copyright, Designs and Patents Act 1988, without the prior permission of the publisher.

First published 2004 by Blackwell Publishing Ltd

Library of Congress Cataloging-in-Publication Data
Harwood, Laurence M.
 Organic chemistry at a glance / Laurence M. Harwood, John E. McKendrick, Roger C. Whitehead.
  p. cm. – (At a glance series)
 Includes index.
 ISBN 0–8654–2782–8 (acid-free paper)
 1. Chemistry, Organic. I. McKendrick, John E. II. Whitehead, Roger C. III. Title.
 IV. At a glance series (Oxford, England)

 QD253.2.H367 2003
 547–dc22
 2003069154

ISBN 0–8654–2782–8

A catalogue record for this title is available from the British Library

Set in 10/12pt Times
by Integra Software Services Pvt. Ltd, Pondicherry, India

The publisher's policy is to use permanent paper from mills that operate a sustainable forestry policy, and which has
been manufactured from pulp processed using acid-free and elementary chlorine-free practices. Furthermore, the
publisher ensures that the text paper and cover board used have met acceptable environmental accreditation standards.

For further information on Blackwell Publishing, visit our website:
www.blackwellpublishing.com

FSC
Mixed Sources
Product group from well-managed
forests and other controlled sources
Cert no. SGS-COC-2953
www.fsc.org
© 1996 Forest Stewardship Council

Contents

Preface

The stimulus for writing this book came from several perceptions of need. The ever widening range of entrants to read chemistry at an advanced level, bringing with them their increasingly varied, yet equally valid, educational pathways prior to entry, raises the challenge of presenting a level playing field for all at the outset of this new educational experience. In addition to being a mainstream subject, organic chemistry plays an important role in many areas of science, from physics and material science to biochemistry and medicine, where the ultimate practitioners need a fundamental understanding of the basics of the subject without the necessity – or desire – to follow it to its most profound limits. Throughout this range of potential students there is a growing pressure to learn efficiently and unambiguously within increasing time constraints. This book attempts to address those needs, to be a source of primary information of the majority of basic organic chemistry and to act as a valuable *aide memoire* for those needing to dip into a subject area, to refresh memory or carry out a little last minute revision.

The aim of this book is to be a pragmatic but effective means of information transfer and to provide basic information; enough to satisfy some yet excite and enthuse others in the increasingly information-rich, time-poor environment of the university arena. To achieve this goal we have opted for a presentation based on images juxtaposed with bullet point statements. The purist may well care to look away, but it is easy to underestimate the academic challenge posed by the very exacting demands of reconciling the delivery of a clear, unambiguous, yet concise message. It is true that, in the far flung reaches of chemical esoterica, there may well be contrived exceptions to certain statements in this book but we have striven to introduce a complex and diverse subject in the most direct and categorical way we could in order to give a clear message devoid of confusion.

We have attempted to cover all areas of major interest for those in their first years studying organic chemistry. Some subjects will surely prove unnecessary for some readers whereas, inevitably in a book of this size, some subjects have been excluded. Spectroscopic analysis stands out as one such omission, but we felt that it could be treated in such a 'stand alone' way that we were justified in omitting it as there are many other works which do justice to this important area of organic chemistry.

The main body of the book is grouped into sections devoted to atomic structure, bonding and molecular structure, and reaction types. Within each double page spread we have attempted to address a coherent range of facts with cross-referencing where relevant so that opening the book anywhere leads the reader to a self-contained choice of subject matter, yet directs to wider ramifications within the whole field of organic chemistry. The last section, compound classes, departs from the standard style and uses flow diagrams to encapsulate the more important functional group interconversions.

We sincerely hope that presenting *Organic Chemistry at a Glance* in this style will facilitate the introduction and consolidation of a fascinating, varied and infinitely complex subject at a level which will satisfy both those who wish to expand their horizons in other areas as well as those who wish to explore this area more deeply.

Laurence M. Harwood, John E. McKendrick, Roger C. Whitehead

1

Atomic Structure

Ground state electronic configuration of carbon

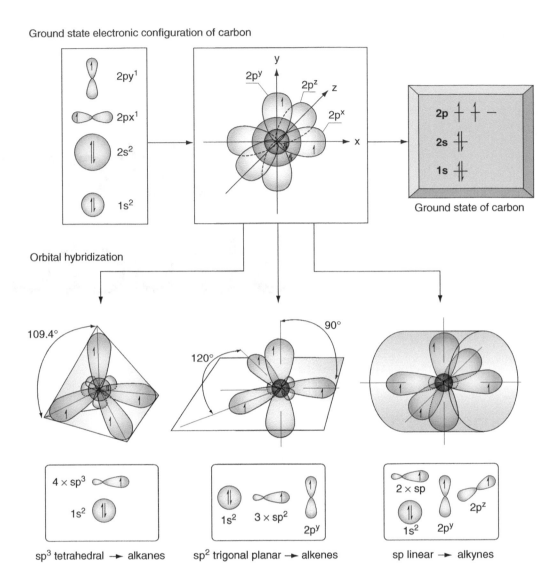

Ground state of carbon

Orbital hybridization

sp³ tetrahedral ➤ alkanes   sp² trigonal planar ➤ alkenes   sp linear ➤ alkynes

1.1 Ground state electronic configuration of atomic carbon

Electrons in an atom occur in *orbitals*. An orbital is a diffuse region around the nucleus where there is a finite probability of finding an electron. The carbon atom in the ground state has six electrons occupying *s-orbitals* and *p-orbitals*.

- *s-orbitals* are spherical with the nucleus at the centre.
- *p-orbitals* are dumb-bell shaped and consist of three orthogonal, equal energy orbitals.

Orbitals are arranged within defined volumes around the nucleus called *shells*. Shells further from the nucleus contain an increasing number of orbitals and have a greater capacity for electrons. The first shell consists of a 1s-orbital containing two electrons; the second shell is made up of one 2s-orbital and three 2p-orbitals which have a combined capacity for eight electrons.

The *ground state electronic configuration* of an atom describes the arrangement of electrons in the lowest-energy state and is assigned by following three simple rules:

- The lowest-energy orbital is filled first (*The Aufbau principle*).
- Each orbital can accommodate two *spin-paired* (↑↓) electrons (*Pauli exclusion principle*).
- Orbitals of equal energy successively accept a single electron until all are half full (*Hund's rule*).

Applying these rules to carbon, the two lowest-energy electrons are placed in a 1s-orbital, the next two electrons are placed in a 2s-orbital and the final two electrons are placed with parallel spins (↑↑) in two of the three available 2p-orbitals ($2p_x^1$, $2p_y^1$).

- The ground state electronic configuration of a carbon atom is $1s^2\, 2s^2\, 2p^2$

1.2 Orbital hybridization

When carbon forms bonds, a 2s electron is promoted into the vacant p-orbital ($2p_z$) giving carbon an electronic configuration of $1s^2\, 2s^1\, 2p^3$. The half-filled 2s- and 2p-orbitals, containing a total of four valence electrons, may combine in three different ways to form *hybrid valence orbitals* which result in different bonding arrangements.

Combination of the 2s-orbital and all three 2p-orbitals gives four equivalent *sp³ hybrid atomic orbitals* pointing towards the corners of a tetrahedron. This situation occurs in *alkanes*.

Combination of the 2s-orbital and two 2p-orbitals gives three equivalent *sp² hybrid atomic orbitals* lying in a plane at 120° to each other and a residual unhybridized 2p-orbital which lies above and below the plane. This situation occurs in *alkenes*.

Combination of the 2s-orbital and one 2p-orbital gives two equivalent *sp hybrid atomic orbitals* disposed linearly and two residual unhybridized 2p-orbitals orthogonal to the sp axis. This situation occurs in *alkynes*.

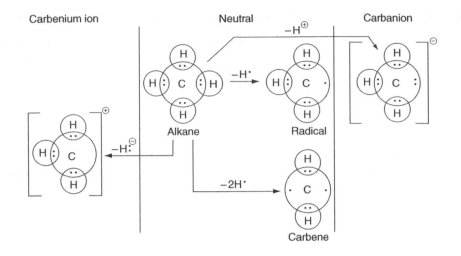

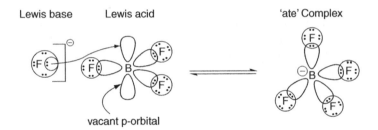

1.3   Electron accounting: the octet rule

The outer shell of carbon with its four valence electrons can achieve the *inert gas closed shell configuration* of eight electrons (*octet rule*) by forming four *covalent bonds* each containing two electrons in a *molecular orbital* shared between the bonding partners. Any charge associated with the carbon may be deduced by the following electron accounting procedure:

- Each bonded atom is formally assigned one of the bonding electrons.
- Non-bonding electrons are assigned fully to the atom with which they are associated.
- Comparison of the number of electrons so assigned with the number of valence electrons possessed by the neutral atom allows calculation of the residual charge.

Trivalent carbon with six bonding electrons in its valence shell has a residual positive charge (*carbocation* or *carbenium ion*).

Divalent carbon with six electrons in its valence shell (*carbene*), trivalent carbon with seven electrons in its valence shell (*radical*) and tetravalent carbon with eight electrons in its valence shell are all neutral species.

Trivalent carbon with eight electrons in its valence shell has a residual negative charge (*carbanion*).

1.4   Lewis acids and Lewis bases

A *Lewis acid* is a species which can accept an electron pair. The Lewis acidic centre has vacancy in the valence shell to accept this pair of electrons. For example, compounds of Group IIIA elements such as BF_3 and $AlCl_3$ achieve a closed shell configuration by accepting an electron pair and are particularly efficient Lewis acids.

A *Lewis base* is an electron-rich species which can donate an electron pair e.g. NH_3.

1.5   Electrophiles and nucleophiles

The terms *electrophile* and *nucleophile* are generally used within the context of bond forming reactions to carbon.

An *electrophile* is an electron-deficient species and will react by accepting electrons in order to attain a filled valence shell. An electrophilic atom may either be a positively charged species such as a *cation* or a neutral species such as a *Lewis acid, carbene* or an electron-deficient radical. An electrophilic centre may also be induced by the presence of a neighbouring electronegative atom; for example, CH_3–I, R_2C=O (see Sections 2.11 and 5.36).

Note: The charged atom of a positively charged species may not always be electrophilic. For example, the carbon atom in CH_5^+ or the oxygen atoms in H_3O^+ and Et_3O^+ possess filled valence shells and, in these cases, the electrophilic site is adjacent to the charged atom.

A *nucleophile* has electrons available for donation to electron-deficient centres. A nucleophilic atom donates two electrons if it is either negatively charged (*anion*), or neutral, but carrying a non-bonding pair of electrons (for example, RNH_2, ROH, H_2O). Nucleophilic radicals also exist which, although formally neutral, react preferentially with electron-deficient species (see Section 5.16).

2

Bonding and Molecular Structure

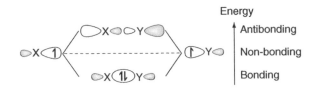

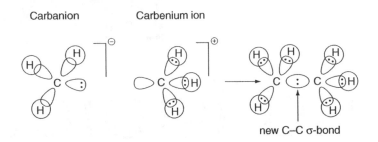

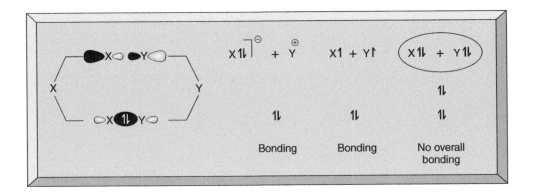

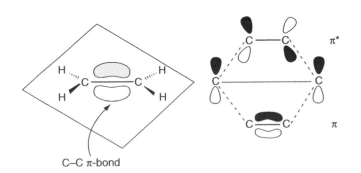

2.1 Bonding and anti-bonding orbitals

Overlap of two *atomic orbitals* gives rise to two *molecular orbitals* known as the *bonding orbital* and *anti-bonding orbital* respectively. Although the overall orbital energy does not alter, the bonding orbital is lower in energy and the anti-bonding orbital correspondingly higher in energy than the original atomic orbitals. As with atomic orbitals (see Section 1.1), molecular orbitals are filled in order of increasing energy and each can contain a maximum of two spin-paired electrons. Therefore, interaction of two half-filled atomic orbitals or a filled and an empty atomic orbital leads to a filled bonding molecular orbital and an empty anti-bonding molecular orbital. This new arrangement is more stable than that of the isolated systems.

A qualitative consideration of orbitals is generally sufficient for organic chemists. A more rigorous mathematical treatment is known as *Linear Combination of Atomic Orbital (LCAO) theory* and leads to the following important conclusions:

- n atomic orbitals combine to give n/2 bonding molecular orbitals and n/2 anti-bonding molecular orbitals.
- The electrons occupying the original atomic orbitals fill the molecular orbitals in order of increasing energy.

2.2 The σ-bond

A σ-bond (σ = sigma) is formed by linear overlap of two atomic orbitals which results in a region of high electron density of circular cross-section concentrated between the two participating nuclei. It is this region of negative charge which attracts and binds the two positively charged nuclei together, overcoming their mutual electrostatic repulsion.

The associated unfilled σ-antibonding molecular orbital has defined regions, or *lobes*, which point away from the two bonding nuclei. If this orbital were to contain electrons, the electron density would be directed linearly away from the σ-bond with no electron density between the nuclei, a region referred to as a *node*.

2.3 The π-bond

A π-bond (π = pi) is formed by sideways overlap of two atomic p-orbitals. Due to the dumb-bell shape of the p-orbitals (see Section 1.1), the regions of high electron density are found as banana-shaped regions above and below a plane containing the two atoms, with no electron density in the plane.

The π-antibonding molecular orbital also has a node in the plane containing the atoms but has an additional node between the atoms orthogonal to that plane. It consists of four lobes directed towards the corners of a rectangle in the same plane as the π-bonding orbital.

Alkanes

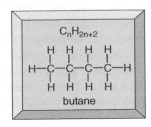

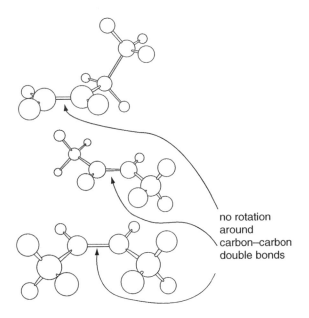

Free rotation possible around all carbon–carbon single bonds

Alkenes

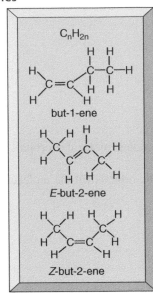

no rotation around carbon–carbon double bonds

Alkynes

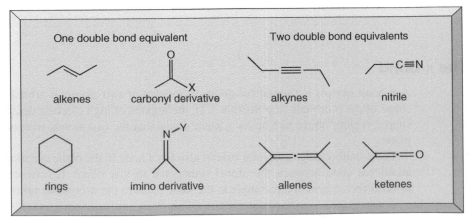

the alkyne unit is linear

One double bond equivalent		Two double bond equivalents	
alkenes	carbonyl derivative	alkynes	nitrile
rings	imino derivative	allenes	ketenes

2.4 Bonding and structure of alkanes, alkenes and alkynes

Alkanes

These are often referred to as *saturated hydrocarbons* because they will not absorb hydrogen. They are composed of sp^3-hybridized carbons (see Section 1.2) attached to each other by σ-bonds (see Section 2.2).

Alkanes may be *acyclic* or *cyclic* structures. Acyclic alkanes have the general formula C_nH_{2n+2} whereas monocyclic alkanes have the general formula C_nH_{2n}. Free rotation is possible about the C–C σ-bonds of acyclic alkanes and so these alkanes can adopt an infinite number of three-dimensional arrangements known as *conformers* (see Section 3.10).

The lack of polarization in the carbon–hydrogen framework of alkanes makes them unreactive towards heterolytic reactions (see Section 2.7) but they will undergo homolytic reactions resulting in substitution of hydrogen atoms (see Sections 2.7 and 5.16).

Alkenes

These possess a double bond between two sp^2-hybridized carbon atoms (see Section 2.3) and are often said to be *unsaturated* as each double bond has the capacity to absorb a further molecule of hydrogen to form an alkane.

Acyclic alkenes possessing one double bond have the general formula C_nH_{2n} and any additional double bond (*degree of unsaturation*) results in loss of two further hydrogens from the general formula. The double bond is composed of a σ-bond (see Section 2.2) and a π-bond (see Section 2.3). Due to the requirement for overlap of the p-atomic orbitals on each carbon atom to form the π-molecular orbital, free rotation around the double bond is not possible. The four substituents on the alkene carbon atoms lie in a plane and are fixed in space, leading to the possibility of *isomers* (see Section 3.6).

The double bond results in high electron density above and below the plane containing the four substituents (see Section 2.3) and this means that alkenes react readily with *electrophiles* (see Section 1.5).

Alkynes

These possess a triple bond between two sp-hybridized carbon atoms (see Section 1.2). Each triple bond has the capacity to absorb two molecules of hydrogen to form an alkane. Absorption of one molecule of hydrogen leads to an alkene.

Acyclic alkynes possessing one triple bond have the general formula C_nH_{2n-2}. The triple bond is composed of a σ-bond and two orthogonal π-bonds. The two remaining substituents on the carbon atoms are arranged linearly.

2.5 Double bond equivalents

The number of *double bond equivalents* in a hydrocarbon is calculated by comparing the number of hydrogen atoms present in the molecule with the number possessed by the corresponding saturated hydrocarbon and dividing the difference by two. From the considerations above, double bond equivalents result from the presence of a double bond, a triple bond or a ring in a molecule.

This concept can be extended to any organic molecule by reducing the structure to that of its *equivalent hydrocarbon* for comparison with the corresponding saturated hydrocarbon. This is derived as follows:

- Replace all monovalent substituents with a hydrogen e.g. F, Cl, Br, I.
- Remove all divalent substituents e.g. O, S.
- Remove all trivalent substituents together with one hydrogen e.g. N.
- The result gives the equivalent hydrocarbon which can be used to calculate the number of double bond equivalents as above.

Note: Double bond equivalents result from structures such as C=O and C≡N as well as C=C and C≡C.

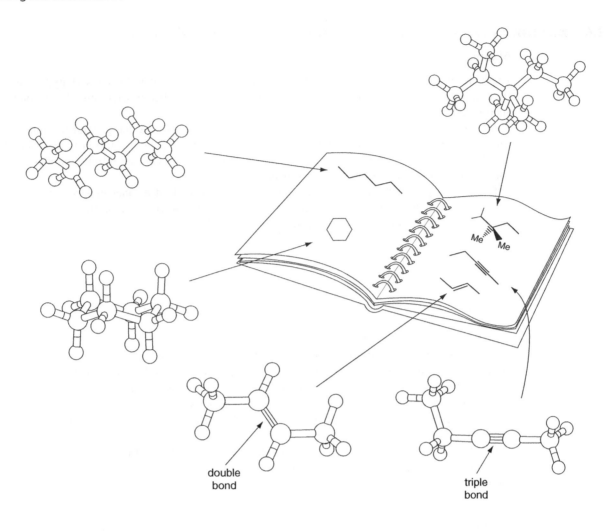

double
bond

triple
bond

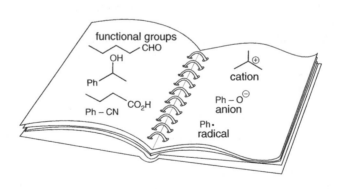

functional groups

cation

anion

radical

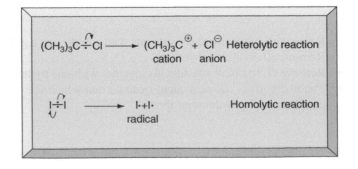

2.6 Drawing molecular structures

For speed and convenience, it is common to represent molecular structure by a shorthand notation which emphasizes *bonds* and *bond angles* in the molecule rather than concentrating solely upon constituent *atoms*. This is because, as chemists, we are mainly concerned with chemical reactions and these involve bond formation and cleavage.

By convention, a bond is represented by a single line and atomic centres are defined to be at any junction between two lines and at the termini. If the atom is carbon, no symbol is employed and any unsatisfied valency is judged to be a hydrogen substituent. All atoms other than carbon or hydrogen are labelled with the corresponding atomic symbol and any substituents they possess.

A chain of sp^3 carbon atoms is drawn as an extended zig-zag line, representing atoms in the plane of the paper. Cyclic structures are depicted by regular polygons in which the number of vertices corresponds to the number of atoms in the ring. Additional substituents are represented using thickened and hashed lines implying substituents projecting out of, or going into, the plane of the paper respectively. This formalism reflects the tetrahedral nature of the sp^3 carbon atoms (see Section 1.2).

A double bond is represented by two parallel lines and, in the case of nitrogen and sp^2 hybridized carbon, the positioning of substituent atoms reflects their relative disposition within the molecule (see Section 1.2).

A triple bond is represented by three parallel lines and, in the case of alkynes, the additional substituent is oriented linearly to reflect the direction of bonding in sp-hybridized carbon (see Section 1.2).

The shorthand notation may be reinforced by replacing commonly encountered part structures by symbols (for example Me=methyl, Et=ethyl, Ph=phenyl, CHO=aldehyde, CO_2H=carboxylic acid, CN=nitrile).

Charges are located on the atomic centres where they are found and are best represented by the charge sign enclosed in a circle. Unpaired electrons on an atom are depicted as a single dot and lone pairs as two dots. While charge and single electrons must be shown, lone pairs are frequently omitted unless involved in a reaction mechanism.

2.7 Depicting electron movement in reaction mechanisms

Two electrons are involved in bond formation or cleavage. They may move as a pair in *heterolytic reactions* or singly in *homolytic reactions*. Heterolytic reactions involve charged species (*cations* and *anions*); whereas homolytic reactions involve species possessing a single unpaired electron (*radicals*).

By convention, chemists depict the movement of a pair of electrons with a double-headed *curly arrow*. The arrow is drawn with its origin at the site of electron density (bond or atom) and is directed to the electrophilic reactive centre (bond or atom). *Remember that every curly arrow represents movement of two spin-paired electrons*.

The movement of a single electron is depicted by a single-headed *curly fish hook*. The making or breaking of a bond, which involves two electrons, requires two curly fish hooks to be drawn, each with its origin at the source of electrons and directed to the electron receiving site.

Any charge resulting from movement of electrons can be assessed by electron accounting (see Section 1.3).

Nomenclature priority

Structure	Name
$-\overset{\oplus}{N}R_3$	Onium ion
—COOH	Carboxylic acid
—SO₃H	Sulfonic acid
—COX	Acid halide
—CONR₂	Amide
—CN	Nitrile
—CHO	Aldehyde
—CO—	Ketone
ROH	Alcohol
ArOH	Phenol
—SH	Thiol
—NR₂	Amine
═	Alkene
≡	Alkyne
R— X—	Other substituents

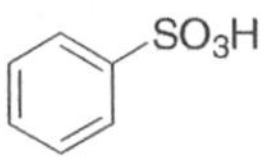

Benzenesulfonic acid

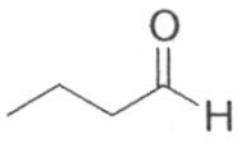

Butanal

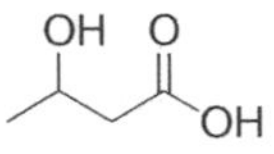

3-Hydroxybutanoic acid

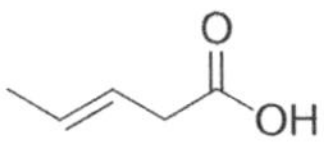

E-Pent-3-enoic acid

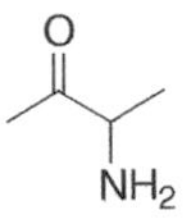

3-Aminobutan-2-one

Choosing the parent name

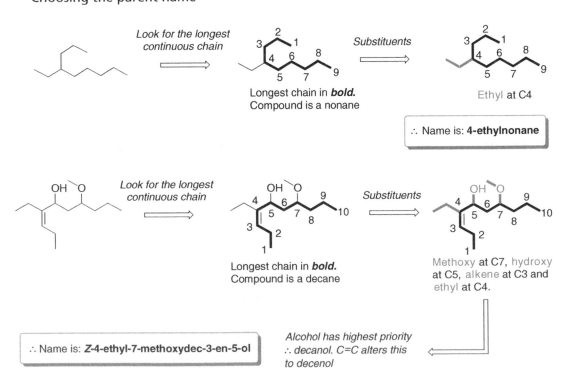

Look for the longest continuous chain ⟹ Longest chain in **bold.**
Compound is a nonane

Substituents ⟹ Ethyl at C4

∴ Name is: **4-ethylnonane**

Look for the longest continuous chain ⟹ Longest chain in **bold.**
Compound is a decane

Substituents ⟹ Methoxy at C7, hydroxy at C5, alkene at C3 and ethyl at C4.

Alcohol has highest priority ∴ *decanol. C=C alters this to decenol*

∴ Name is: **Z-4-ethyl-7-methoxydec-3-en-5-ol**

INTRODUCTION TO NOMENCLATURE

2.8 General nomenclature punctuation

> • Except for carboxylic acids, their derivatives and *N*-alkyl amines, compound names are almost invariably one continuous word.
> • Numbers are separated from letters by '-'.
> • Numbers are separated from numbers by ',' (except in polycycles).
> • Substituents are ordered alphabetically, regardless of their priority.
> • Bracket hierarchy is {[()]}. Side chain numbering hierarchy is 1', 2'..., 1", 2"..., 1''', 2'''...etc.

2.9 Determining the primary functionality

Priority is determined by the atomic number of the atoms constituting the functionality, multiple bonded atoms taking priority over singly bonded atoms. The thiol group is the exception, being ranked as an alcohol.

Functional groups which are included in the carbon atom of the stem (see below) are more important than substituents on the stem. Such primary substituents are placed at the end of the name. The exceptions are alkene and alkyne substituents which remain within the stem, with *yne* being ranked before *ene*.

> $-CO_2R$ (*stemoic acid, stemoate*) ≥ >C=O (*stemone*) > –CHO (*stemal*) > –OH (*stemol*) + –SH (*stemthiol*) > –C≡C– (*stemyne*) ≥ >C=C< (*stemene*) > $-NR_2$ (*stemamine*) > –CN (*stemonitrile*)

Other substituents are expressed as a prefix to the stem:

> –I (*iodostem*) ≥ –Br (*bromostem*) ≥ –Cl (*chlorostem*) >
> –F (*fluorostem*) > –OR (*alkoxystem*) ≥ $-NO_2$ (*nitrostem*) etc.

When there is more than one functionality forming part of the stem, subordinate functionalities are named as prefixes (except for double and triple bonds which are retained in the stem in order of priority):

> >C=O, –CHO (*oxostem*) > –OH (*hydroxystem*) > $-NR_2$ (*aminostem*) > –CN (*cyanostem*) > –R (*alkylstem*)

2.10 Choosing and numbering the parent name (stem)

Acyclic systems

Choose the longest straight chain containing the primary functionality and number it from the end which places the primary functionality at the lowest position on the chain. Do not forget to include carbonyl carbons as part of the chain. Chains up to four carbons long take trivial names, longer chains are named according to chain length. If the chain is saturated the stem is named as an *alkane*, if it contains unsaturation it is named as an *alkene* or an *alkyne*. Alkenes may be *E*- or *Z*- (see Section 3.6). The stem loses the final letter if it comes before a functionality beginning with a consonant.

Number of carbons:	1	2	3	4	5	6	7	8	9	10
stemane (*yne, ene*)	meth	eth	prop	but	pent	hex	hept	oct	non	dec

The number locating the position of the substituent on the chain precedes its alphabetically determined placement in the complete name.

Alkyl substituents which are themselves substituted are numbered following the same procedure, except that numbering starts at the carbon attached to the main stem. The subordinate chains are enclosed in brackets and the numbering may be labelled as 1' – n', 1" – n" to imply increasing subordinacy of different substituent chains.

Alkyl substituents attached by a carbonyl carbon are termed *stemoyl*.

Monocyclic aliphatic systems

| Cyclopentane | Cyclopropane | Cyclopentane carboxylic acid | Cyclohexane carboxaldehyde | Cyclobutanol | 3-Methylcyclohexanol |

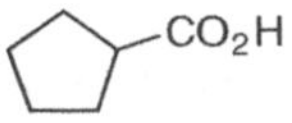

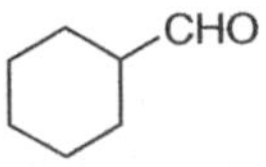

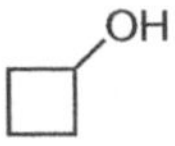

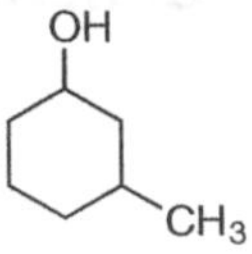

Aromatic systems

| 1,2- or ortho- or o- | 1,2-Dibromobenzene | 1,3- or meta- or m- | meta-Chlorotoluene | 1,4- or para- or p- | p-Methylphenol |

| Benzoic acid | Aniline | Toluene | Phenol | Pyrrole | Furan | Thiophene |

| Pyridine | Pyrylium | Indole | Benzofuran | Quinoline | Isoquinoline |

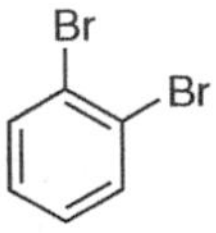

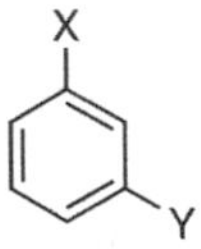

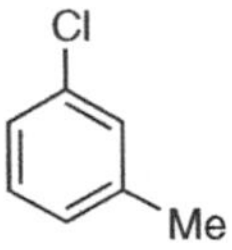

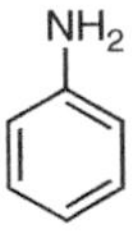

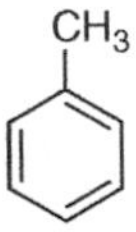

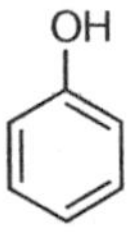

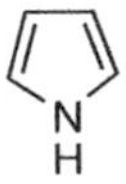

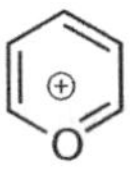

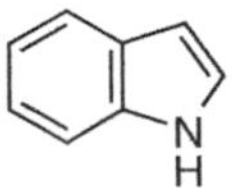

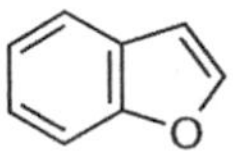

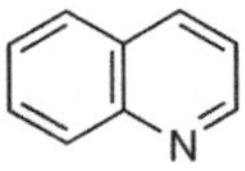

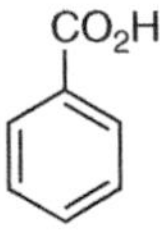

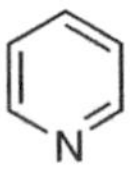

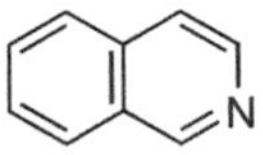

Polycyclic systems

Bicyclo[2.2.1]heptane

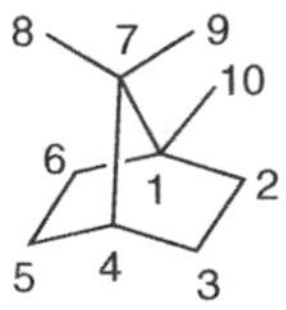

1,7,7-Trimethylbicyclo[2.2.1]heptane

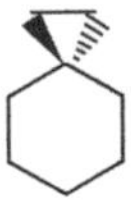

Spiro[2.5]octane

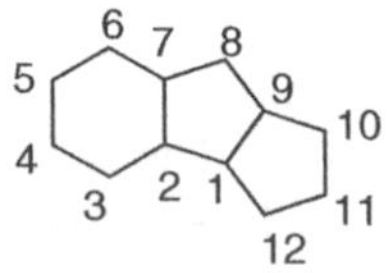

Tricyclo[7.3.0^{2,7}]dodecane

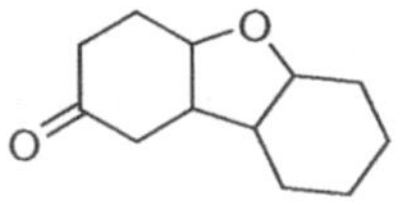

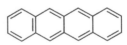

8-Oxatricyclo[7.4.0^{2,7}]tridecan-4-one

Tetracene

Monocyclic aliphatic systems

The ring is named after the number of carbons contained within it with the prefix *cyclo*. The primary functionality is placed at C-1 by convention and its position does not appear in the compound name. The ring is numbered in the direction which places the next highest priority substituent at the lowest numbered position.

If a carboxylic acid is attached directly to the ring, the compound is named as a *cycloalkyl carboxylic acid*. An aldehyde group attached to the ring is named as *cycloalkyl carboxaldehyde*. *Note*: These are examples where the name is broken into two words.

All other rules follow those for acyclic aliphatic substrates.

Aromatic systems

Substituted benzenes are numbered as aliphatic cyclic systems. All other rules follow those for cyclic aliphatic compounds. Disubstituted systems are still often referred to by the old, trivial terminology:

1,2-disubstituted *ortho*; 1,3-disubstituted *meta*; 1,4-disubstituted *para*

IUPAC rules allow certain trivial names to be used:

benzene carboxylic acid : benzoic acid	aminobenzene : aniline
hydroxybenzene : phenol	methylbenzene : toluene

Aromatic heterocycles have specific names; commonly encountered heterocycles are listed:

heteroatom	nitrogen	oxygen	sulfur
5-membered ring	pyrrole	furan	thiophene
benzannelated derivative	indole	benzofuran	
6-membered ring	pyridine	pyrylium	
benzannelated derivative	quinoline, isoquinoline		

With the exception of isoquinoline, which is numbered in the same way as quinoline, numbering starts at the heteroatom. In the case of the benzannelated derivatives, the hetero-ring is numbered first and atoms common to both rings are numbered last.

Polycyclic systems

A polycyclic system is considered to have the same number of rings as the number of bond cleavages required to reduce it to an acyclic structure.

A bicyclic structure having three bridging chains joined at *two* common **bridgehead** carbon atoms is named as a ***bicyclo[a.b.c]alkane***. Here a, b and c refer to the number of atoms in each bridging chain, ordered according to chain length and the alkane corresponds to the ***total*** number of atoms in the skeleton. Numbering of the system starts at a bridgehead atom, proceeds around the longest bridge to the second bridgehead atom, returns *via* the next longest chain to the original bridgehead and then finishes on the shortest chain, counting from the first bridgehead. The C-1 bridgehead is chosen to minimize the position of the highest priority substituent of the longest substituted chain. *Note*: Figures indicating bridging chain lengths are separated by points, not commas.

A bicycle possessing two bridging chains joined at a *single* common atom is termed a ***spiro[a.b]alkane*** (a combination of Latin *spiro* and Greek *spira* meaning twisted) where a and b again refer to the number of atoms in the bridges. Confusingly, in such cases the chains are exceptionally ordered in ***increasing*** length.

All other polycyclic systems are named by first identifying the longest possible bridging chain. This then defines the bridgehead carbons and the remaining two chains bridging the bridgehead atoms are listed according to size. All other subordinate bridging chains are listed subsequently with their points of attachment to the main chains being indicated by superscripts such as ***tricyclo[a.b.c.d^{x,y}]alkane*** etc.

Polycyclic systems containing heteroatoms in the skeleton are named as the carbon analogue with prefixes n-oxa, n-aza, n-thia etc. where n represents the position of the heteroatom in the skeleton.

Bond polarization

$$H_3C\text{—}CH_3$$

no polarization of the C–C bond.

$$\overset{\delta+}{H C_3}\text{—}\overset{\delta-}{C C l_3}$$

Cl relative to C is electronegative, this causes polarization of the C–C bond as shown.

Br and O are electronegative, ∴ bond is polarized.

C=O bond exhibits the same polarization as the C–O bond.

Electronegativity on the Pauling scale

			H		
			2.1		
Li		B	C	N	O	F
1.0		2.0	2.5	3.0	3.5	4.0
	Mg	Al	Si	P	S	Cl
	1.2	1.5	1.8	2.1	2.5	3.0

Resonance

Extreme canonicals

Disfavoured, however still CONTRIBUTES to overall electronic structure.

More favourable as O bears a negative charge, more important contribution to overall electronic structure.

All p-orbitals are aligned. Electrons can move through the aligned orbitals

Anions and cations may be stabilized by resonance

Double Bond

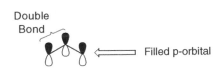 ⟸ Filled p-orbital

Orbitals are aligned, charge can be shared out through the π-system.

Double Bond

 ⟸ Empty p-orbital

2.11 Bond polarization

A bond between two atoms contains two electrons occupying the bonding molecular orbital (see Sections 2.1 and 2.2). If the two atoms are the same, the overall electron distribution is symmetrical about a plane bisecting the bond equidistant between the two atoms. Such a bond is said to be **unpolarized** or **non-polar**.

The two bonding atoms need not be the same (e.g. C–O, C–Br) or, if they are the same, may possess different substituents (e.g. Cl_3C–CH_3). These situations result in distortion of the electron cloud with the greater share of electrons going to the more **electronegative** partner. Such a bond is said to be **polarized**, although it remains predominantly covalent. The magnitude of distortion depends upon the difference in electronegativity of the bonding partners. Conventionally the direction of polarization is denoted by an arrow on the bond, with δ^+, δ^- signs on the corresponding atoms. The magnitude of bond polarization diminishes rapidly with increasing distance from the substituent (**inductive effect**).

2.12 Resonance

Any number of adjacent p-atomic orbitals may interact in the same way that two adjacent p-atomic orbitals form bonding and antibonding π-molecular orbitals (see Section 2.3). In this case, the π-electrons are no longer confined to a region between two specific atoms but are **delocalized** along the whole length of the π-system. This effect is called **resonance** and such systems are said to be **conjugated**.

We represent delocalization within a molecule by drawing dashed bonds to indicate partial π-bonds and such a structure is referred as a **resonance hybrid**. This can be considered to be the result of a combination of a number of bond-localized structures referred to as **resonance** or **canonical forms**.

- Extreme resonance canonical forms **are not** distinct entities but represent contributing electronic distributions within the π-framework of the molecule.
- Only electrons move in resonance, the atoms remain fixed in space.

The inter-relationship of canonical forms is conventionally depicted by a double-headed arrow ($\leftrightarrow$). Canonicals have the advantage that they are easier to draw, provide insight into the nature of π-electron distribution, and simplify representation of electron movement in reaction mechanisms compared with the resonance hybrid structure representation.

Not all canonicals contribute equally to the overall resonance hybrid structure and some are more energetically reasonable than others. For instance, charge-separated canonicals of a neutral molecule are highly disfavoured; whereas canonicals of charged molecules in which electron density is localized on an atom of complementary electronegativity are important contributors.

- When drawing different canonical forms it is recommended that you show the movement of the electrons using curly arrows or fish hooks (see Sections 2.7 and 2.14) and use electron accounting (see Section 1.3) to trace the movement of any charges in the system.

2.13 Dipole moments

The distribution of electronegative and electropositive centres, the presence of lone pairs and the combination of inductive and resonance effects may culminate in a molecule possessing net polarization. The magnitude and direction of this is referred to as the **dipole moment** (μ) and it is important in determining how molecules align and interact.

Benzene

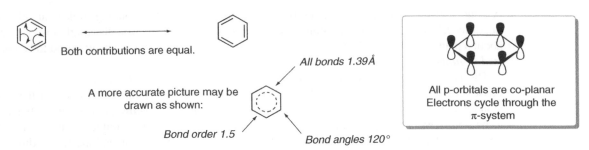

Both contributions are equal.

A more accurate picture may be drawn as shown:

All bonds 1.39Å

Bond order 1.5

Bond angles 120°

All p-orbitals are co-planar
Electrons cycle through the
π-system

Aromatic stability

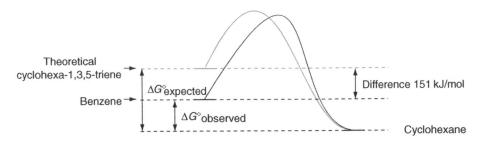

Theoretical cyclohexa-1,3,5-triene

$\Delta G^\circ_{expected}$

Benzene

$\Delta G^\circ_{observed}$

Difference 151 kJ/mol

Cyclohexane

Benzene undergoes substitution reactions with electrophiles, conjugated dienes undergo addition reactions. The stability of the aromatic system causes the difference in reactivity.

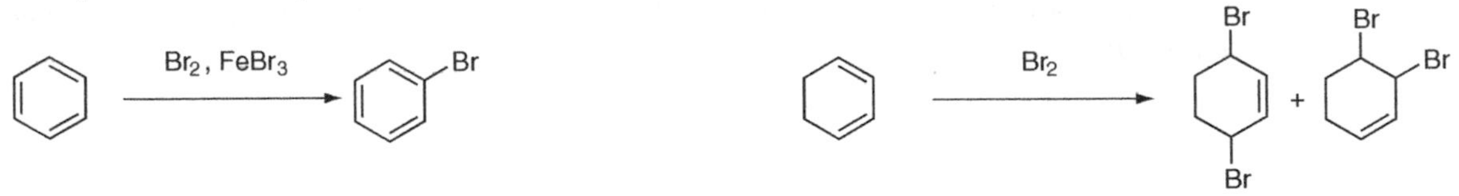

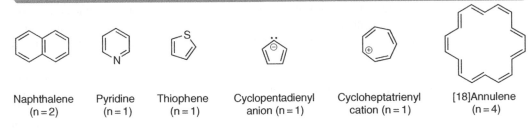

Br_2, $FeBr_3$

Br_2

Hückel's aromatics

Planar fully conjugated monocyclic systems with [4n+2]π electrons have a closed shell of electrons all in bonding orbitals and are exceptionally stable. Such systems are said to be aromatic.

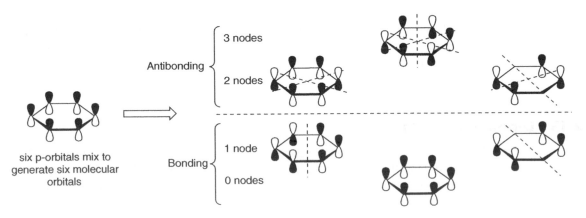

| Naphthalene (n = 2) | Pyridine (n = 1) | Thiophene (n = 1) | Cyclopentadienyl anion (n = 1) | Cycloheptatrienyl cation (n = 1) | [18]Annulene (n = 4) |

π-Molecular orbitals of benzene

six p-orbitals mix to generate six molecular orbitals

Antibonding
- 3 nodes
- 2 nodes

Bonding
- 1 node
- 0 nodes

2.14   Benzene: reactivity and structure

Benzene is the parent member of a series of conjugated unsaturated carbocycles termed *arenes* or *aromatic hydrocarbons*. It has two extreme resonance canonicals, formally represented as cyclohexatrienes. However, benzene displays chemical and physical properties which are not consistent with these bond localized representations. These include:

(1)   The heat of hydrogenation of benzene is significantly less than for 1,3,5-hexatriene or than would be predicted for a theoretical cyclohexatriene.
(2)   Benzene predominantly undergoes substitution reactions with electrophilic reagents (see Section 5.19); whereas conjugated alkenes undergo addition reactions.
(3)   The molecule is planar and all of the C–C bond lengths are identical.

This increased stability is referred to as *aromatic stabilization*.

2.15   Hückel's rule

To exhibit aromatic stabilization, a molecule must possess a cyclic, planar, overlapping array of p-orbitals containing a total of $(4n+2)$ π electrons (where n is any integer).

- As well as including various ring sizes (annulenes), Hückel's rule encompasses polycyclic systems (e.g. naphthalene), rings containing atoms other than carbon (e.g. pyridine) and charged species (e.g. cyclopentadienyl anion).

2.16   The π-molecular orbitals of benzene

The benzene ring is made up of six sp^2-hybridized carbons with internal bond angles of 120°. The carbon–carbon bond lengths are all equivalent (1.39Å) and intermediate between that of a single bond between two sp^2-hybridized carbons (1.47Å) and a carbon–carbon double bond (1.33Å).

The six p-orbitals project above and below the plane of the ring and interact to form six molecular orbitals (three bonding and three antibonding) in an analogous fashion to π-bond formation in ethene (see Section 2.3).

The three bonding π-molecular orbitals are filled by the six π-electrons and the antibonding orbitals are empty. This is responsible for aromatic stability.

Benzene has regions of high π-electron density above and below the ring.

2.17   Drawing aromatic structures and reaction mechanisms

For the purposes of clarity, it is customary to use bond localized *Kekulé structures* (named after the chemist who proposed this structure for benzene in 1866) to depict aromatic substrates in reaction mechanisms. You must remember that the Kekulé forms actually depict resonance canonicals which are freely interchangeable. There is only one 1,2-dibromobenzene for instance. A more precise pictorial representation is a dashed circle inside the six-membered ring to depict the delocalized π-system but this is complicated to use in electron pushing mechanisms (use of an unbroken circle inside a six-membered ring is *incorrect*).

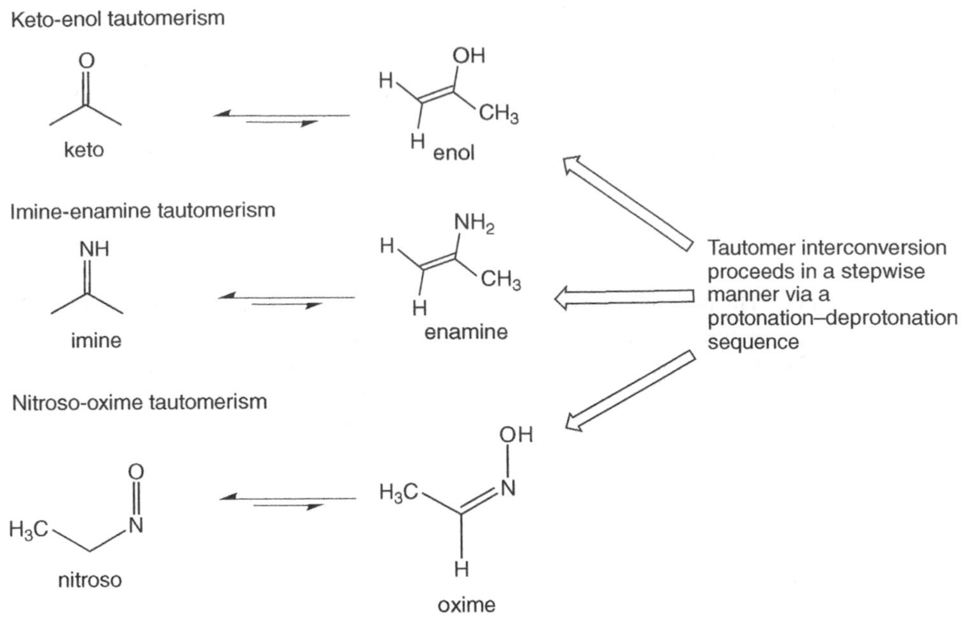

Keto-enol tautomerism

keto ⇌ enol

Imine-enamine tautomerism

imine ⇌ enamine

Nitroso-oxime tautomerism

nitroso ⇌ oxime

Tautomer interconversion proceeds in a stepwise manner via a protonation–deprotonation sequence

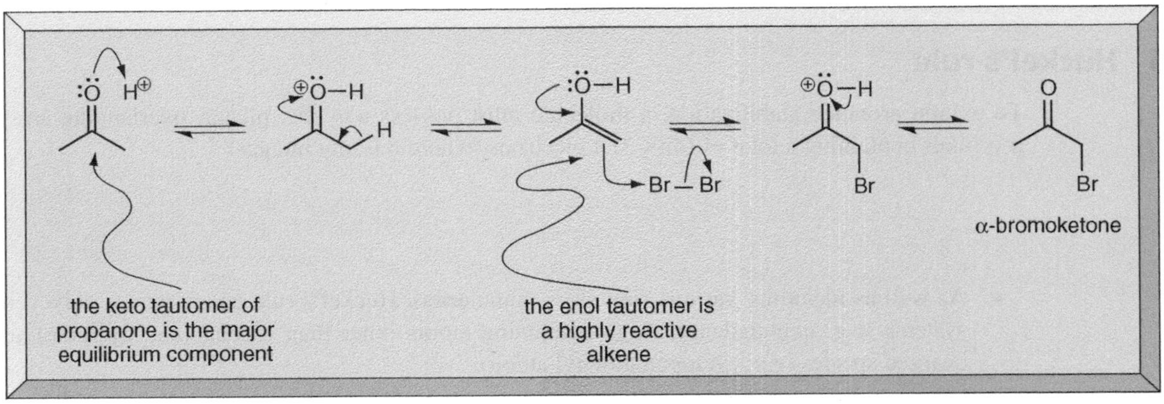

the keto tautomer of propanone is the major equilibrium component

the enol tautomer is a highly reactive alkene

α-bromoketone

Ring-chain tautomerism

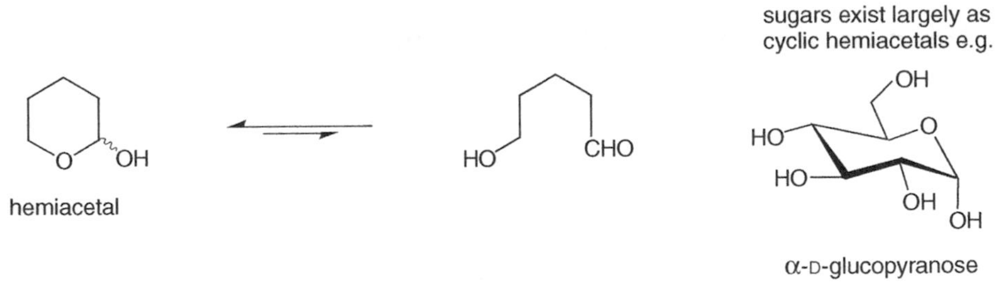

hemiacetal ⇌ HO―CHO

sugars exist largely as cyclic hemiacetals e.g.

α-D-glucopyranose

Valence tautomerism

cycloheptatriene ⇌ norcaradiene   oxepin ⇌ benzene oxide

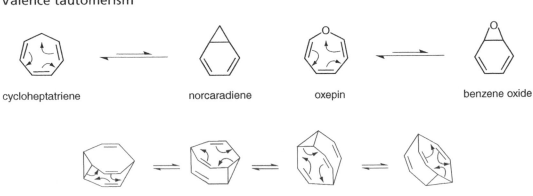

Bullvalene has $\frac{10\,!}{3}$ degenerate valence tautomers

2.18 Tautomers and tautomerism

Tautomers are constitutional isomers (see Section 3.1) which can readily interconvert by transfer of an atom or a group of atoms from one position of the molecule to another and the process is called *tautomerism*.

Keto-enol, imine-enamine and *nitroso-oxime tautomers* interconvert by movement of a proton.

- Keto-enol, imine-enamine and nitroso-oxime tautomerism proceeds by a stepwise protonation–deprotonation pathway.

The relative proportion of tautomers at equilibrium depends upon their structure and external factors such as solvent, pH and temperature. Simple carbonyl compounds exist largely as keto tautomers; whereas 1,3-dicarbonyl compounds may exist substantially in the enolic form. Phenols are extreme cases in which the enol form is strongly favoured due to aromaticity (see Section 2.14).

Tautomers possess different reactivity. Either tautomer may dominate the chemistry of the system regardless of the equilibrium composition. Selective reaction of one tautomer results in its replacement by rapid re-equilibration. For example, the α-bromination of simple ketones under acidic conditions proceeds via the enol despite the fact that this tautomer is usually only present in parts per million.

Ring-chain tautomers are cyclic and acyclic isomers which commonly interconvert by proton transfer. For example, carbonyl compounds possessing a suitably disposed hydroxyl group exist in equilibrium with their cyclic counterparts (*hemiacetals*). This tautomerism is central to the chemistry of sugars which may react selectively via the cyclic or acyclic tautomer.

Valence tautomers interconvert via intramolecular rearrangement and are said to be *fluxional*. Examples include *cycloheptatriene* $\rightleftharpoons$ *norcaradiene* and *benzene oxide* $\rightleftharpoons$ *oxepin*.

- Valence tautomerism involves molecular interconversion and results in movement of atoms. The rate may be altered by external factors. It must not be confused with resonance where only electrons move (see Section 2.12).

Bullvalene has been designed such that rearrangement gives a product with the same structure as the starting material. It can undergo 10!/3 (more than 1.2 million) *degenerate* rearrangements causing all carbons to experience the same environment on average. This applies equally to the hydrogens.

2.19 Hydrogen bonding

A *hydrogen bond* is a dipole–dipole interaction between a hydrogen attached to a small electronegative atom (O, N, F) and the lone pairs of another electronegative atom. It is stronger than simple dipole–dipole interactions (see Section 2.13) (F–H$\cdots$F, 25–30 kJ mol^{-1}; O–H$\cdots$O and N–H$\cdots$N, 12–25 kJ mol^{-1}).

Intermolecular hydrogen bonding is responsible for the fact that hydroxyl-containing compounds generally have boiling points in excess of predictions based on their molecular weights (butan-1-ol 118°C, ethoxyethane 35°C). This effect is far less marked when intramolecular hydrogen bonding is possible.

$$\text{B:} + \text{H–A} \rightleftharpoons \overset{\oplus}{\text{B–H}} + \overset{\ominus}{\text{A}}$$

| Base | Acid | | Conjugate acid | | Conjugate base |

| **Acidity constants for representative Brønsted acids** |

Acid	Conjugate base	Acidity constant K_a	pK_a
HI	$I^{\ominus}$	$ca\ 10^{10}$	$ca\ -10$
HBr	$Br^{\ominus}$	$ca\ 10^{9}$	$ca\ -9$
HCl	$Cl^{\ominus}$	$ca\ 10^{7}$	$ca\ -7$
H_2SO_4	$HO-\overset{O}{\underset{O}{S}}-O^{\ominus}$	1.6×10^{5}	-4.8
$H_3O^{\oplus}$	H_2O	55	-1.7
HF	$F^{\ominus}$	3.5×10^{-4}	3.5
CH_3CO_2H	$CH_3\overset{O^{\ominus}}{\underset{O}{C}}$	1.8×10^{-5}	4.7
$\overset{\oplus}{NH_4}$	$:NH_3$	5.6×10^{-10}	9.2
H_2O	$HO^{\ominus}$	1.8×10^{-16}	15.7^*
CH_3OH	$CH_3O^{\ominus}$	$ca\ 10^{-16}$	$ca\ 16$
$(CH_3)_2CHOH$	$(CH_3)_2CHO^{\ominus}$	$ca\ 10^{-17}$	$ca\ 17$
$(CH_3)_3COH$	$(CH_3)_3CO^{\ominus}$	$ca\ 10^{-18}$	$ca\ 18$
$HC{\equiv}CH$	$HC{\equiv}C^{\ominus}$	$ca\ 10^{-26}$	$ca\ 26$
NH_3	$\overset{\ominus}{NH_2}$	$ca\ 10^{-36}$	$ca\ 36$
$(CH_3)_3CH$	$(CH_3)_3C^{\ominus}$	$ca\ 10^{-71}$	$ca\ 71$

decreasing acidity →

*This value is obtained taking into account that 1 litre of water with a K_a 10^{-14} contains 55.5 moles of water. All other values are for molar solutions.

In the table the conjugate base will deprotonate any acid above it (but also one below it (partially!)).

Base Strength

$$Na^{\oplus}\ \overset{\ominus}{NH_2} + HC{\equiv}CH \longrightarrow HC{\equiv}C^{\ominus}\ Na^{\oplus} + NH_3$$

$$Na^{\oplus}\ \overset{\ominus}{OH} + CH_3CO_2H \longrightarrow CH_3CO_2^{\ominus}\ Na^{\oplus} + H_2O$$

2.20 Acidity constant

Acidity is dependent upon the degree of dissociation of a species HX:

$$HX \rightleftharpoons H^+ + X^-$$

In water, we can represent this equilibrium as:

$$HX + H_2O \rightleftharpoons H_3O^+ + X^-$$

In this equation, H_3O^+ is termed the *conjugate acid* of H_2O and X^- is termed the *conjugate base* of HX.

The *equilibrium constant K* for this is defined by:

$$K = \frac{[H_3O^+][X^-]}{[HX][H_2O]}$$

As water is present in vast excess in this system, its concentration remains effectively constant and can be incorporated into the equilibrium constant to give the *acidity constant K_a*, defined as:

$$K_a = \frac{[H_3O^+][X^-]}{[HX]}$$

Acidity constants can vary by very many orders of magnitude and it is more convenient to use the *pK_a scale* in which *pK_a* is defined by:

$$pK_a = -\log_{10} K_a$$

The *pK_a* values of interest to organic chemists range from -10 (strong acid) to *circa* $+50$.

2.21 Acid strength

The strength of an acid reflects its degree of dissociation which is determined by the net energy gained in the process (*Gibbs Free Energy*, ΔG). This energy change is a consequence of the energy input required to break the HX bond and the energy returned by the formation of the solvated ions, H^+_{aq} and X^-_{aq}. Strong acids are almost completely dissociated.

- Strong acids (HCl, HNO_3, H_2SO_4) have negative *pK_a* values.
- Carboxylic acids have *pK_a* values in the region $0 \rightarrow 5$.

2.22 Base strength

The strength of a base is inversely dependent upon the strength of its conjugate acid. Strong acids have weak conjugate bases; whereas weak acids have strong conjugate bases.

- Strong bases (MeO^-, H_2N^-, R_3C^-) have conjugate acids (MeOH, NH_3, R_3CH) with *pK_a* values greater than 15.
- Weak bases (Cl^-, HSO_4^-, RCO_2^-) have conjugate acids (HCl, H_2SO_4, RCO_2H) with *pK_a* values less than 5.

3

Configurational and Conformational Analysis

Constitutional isomers of pentane

pentane
(n-pentane)

2-methylbutane
(isopentane)

2,2-dimethyl propane
(neopentane)

Chiral objects are non-superimposable mirror images

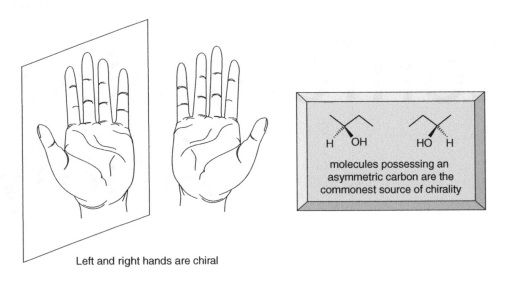

H ⎯ OH   HO ⎯ H

molecules possessing an
asymmetric carbon are the
commonest source of chirality

Left and right hands are chiral

Absolute configuration determination

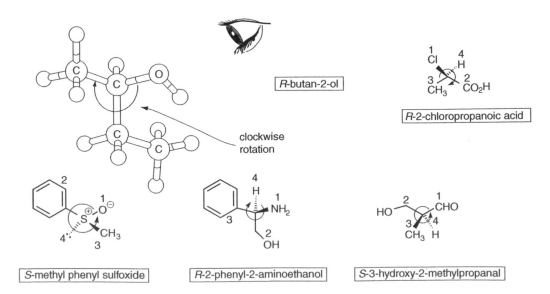

R-butan-2-ol

clockwise
rotation

R-2-chloropropanoic acid

S-methyl phenyl sulfoxide

R-2-phenyl-2-aminoethanol

S-3-hydroxy-2-methylpropanal

3.1 Isomerism

Isomers are molecules which have the same molecular formula but different arrangements of their constituent atoms.

Constitutional isomers (*structural isomers*) have different bonding arrangements of their atoms (*connectivity*) and usually show very marked differences in physical and chemical properties. Connectivity differences can involve the carbon skeleton or the nature and position of functional groups.

Stereoisomers are molecules with identical connectivity but different spatial arrangements of their constituent atoms which cannot be interconverted by bond rotation.

3.2 Sequence rules

In order to categorize stereoisomers it is necessary to prioritize different atomic substituents using the *Cahn–Ingold and Prelog sequence rules* which are applied, in order, until a distinction between substituents is found.

(1) Rank substituents in order of decreasing atomic number of the first bound atom.
Any higher isotope takes precedence over a lower isotope ($^3H > {}^2H > {}^1H$).
A lone pair counts as the lowest priority substituent.
(2) If no distinction is possible at the first atom, consider atoms at increasing distances until a difference is found.
(3) For the purposes of ranking, a multiple-bonded atom is considered equivalent to the same multiple of single-bonded atoms and has higher priority than the corresponding single-bonded substituent.

3.3 Chirality (stereogenicity)

Chirality (*cheir*, Greek for *hand*) refers to objects which are related as non-superimposable mirror images and the term derives from the fact that left and right hands are examples of chiral objects.

sp³-Hybridized carbon atoms possessing four different substituents display this property due to their tetrahedral geometry. Such an asymmetrically substituted carbon atom is a *stereogenic centre* and is the commonest source of chirality in organic molecules.

Unambiguous definition of the spatial arrangement of substituents on a stereogenic centre which distinguishes mirror images gives the *absolute configuration*. A convention permitting structural distinction between two opposite absolute configurations is based upon the sequence rules. It is independent of the chemical or physical properties of the molecule and is equally applicable to tetrahedral stereocentres other than carbon.

(1) Rank substituents on the stereogenic centre in order of decreasing priority using the sequence rules.
(2) View the stereogenic centre with the lowest priority substituent pointing away.
(3) If the order of priority of the three remaining substituents decreases in a clockwise manner the centre is defined as (*R*) – (*rectus*, Latin for *right*). If the order decreases in an anti-clockwise direction the centre is defined as (*S*) – (*sinister*, Latin for *left*).

Enantiomers: non-superimposable mirror images

Chirality due to asymmetric centres

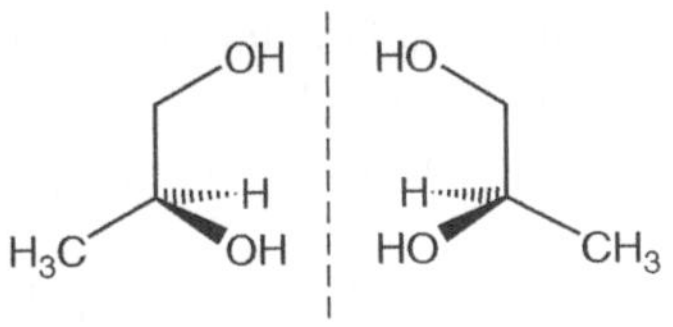

asymmetric carbon 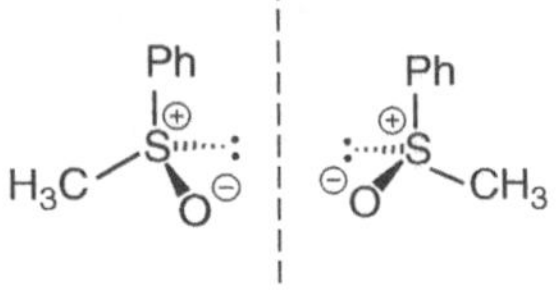   asymmetric sulfur 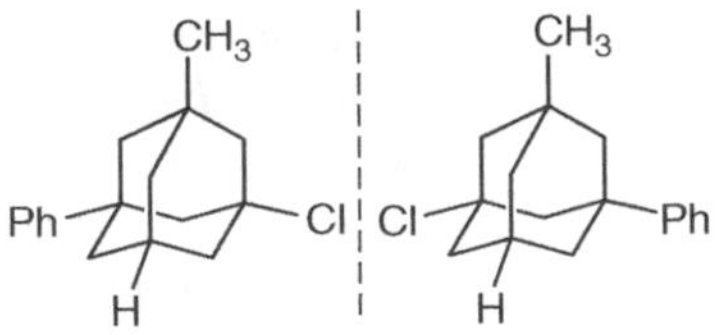   differentially substituted adamantane

Chirality due to orthogonality

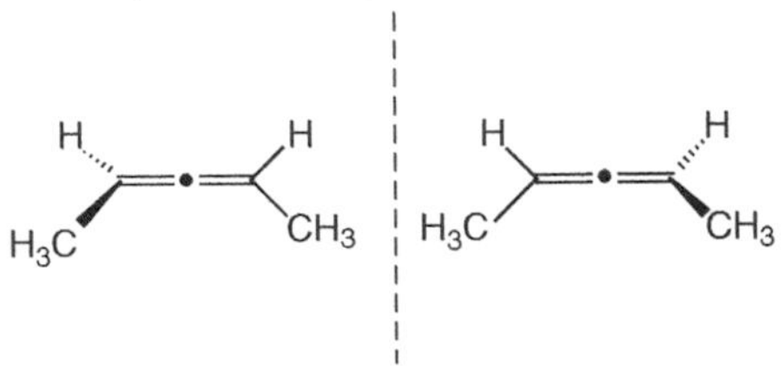

allenes 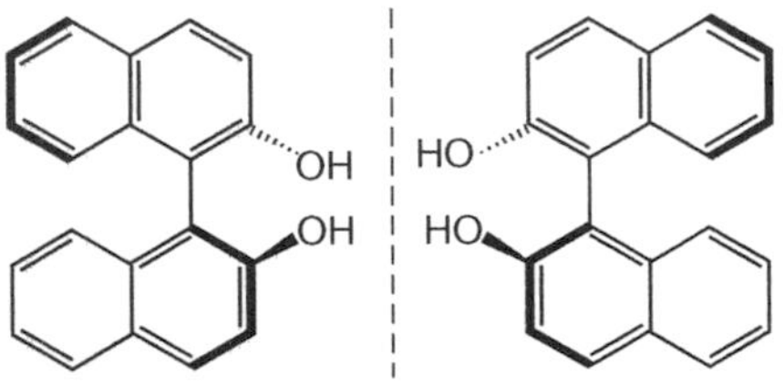   hindered biphenyls

Chirality due to helicity

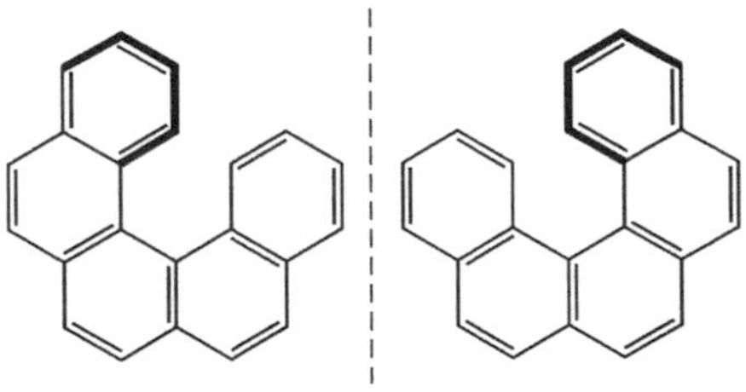

helicenes 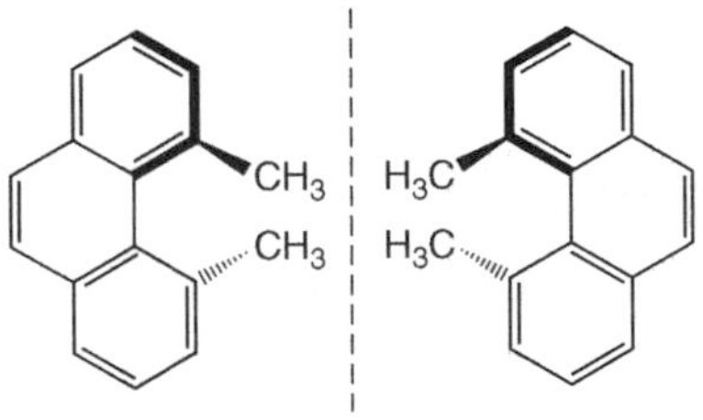   1,8-disubstituted phenanthrenes 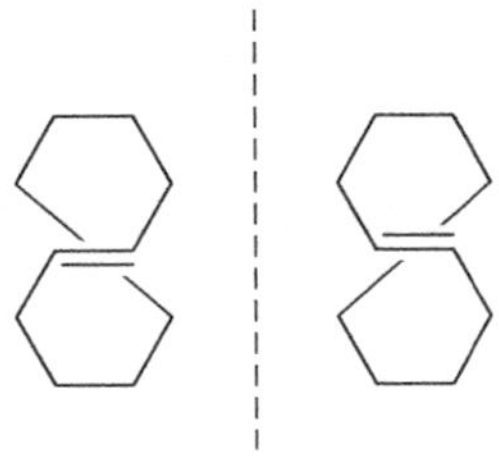   cyclic *E*-alkene

enantiomers rotate plane polarized light in equal but opposite directions

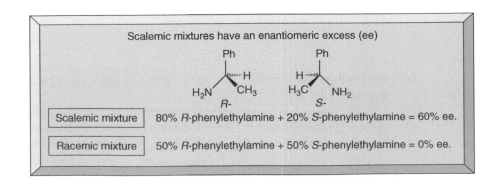

Scalemic mixtures have an enantiomeric excess (ee)

R-   *S*-

Scalemic mixture	80% *R*-phenylethylamine + 20% *S*-phenylethylamine = 60% ee.
Racemic mixture	50% *R*-phenylethylamine + 50% *S*-phenylethylamine = 0% ee.

3.4 Enantiomers

An *enantiomer* is one of a pair of stereoisomers which are related as non-superimposable mirror images. Enantiomerism commonly results from the presence of one or more stereogenic centres in a molecule but may also occur in orthogonal structures (allenes, hindered biaryls), helical structures (*E*-cyclic alkenes, helicenes) and extended tetrahedra (differentially substituted adamantanes). Such molecules are *chiral* and display identical chemical and physical properties in an achiral environment. However, opposite enantiomers will react at different rates with a single enantiomer of a reagent. A single enantiomer will rotate the plane of plane-polarized light and is referred to as *optically active* (see below); although this physical property cannot be directly related to absolute configuration of the molecule. An enantiomer is given the prefix (+)- if the rotation is clockwise and (−)- if the rotation is anticlockwise.

An equal mixture of opposite enantiomers is a *racemate* and solutions of *racemic mixtures* do not rotate the plane of plane-polarized light. Clearly, unequal mixtures of two enantiomers will have a lower optical rotation than a pure enantiomer and the strength of this rotation will depend upon the enantiomeric *excess* (*e.e.*) of the mixture. Mixtures of unequal amounts of enantiomers are referred to as *scalemic*.

Enantiomeric excess = (%age enantiomer A − %age enantiomer B)%

3.5 Specific rotation (α)

The *specific rotation* enables comparison of optical activity between samples by standardizing the analysis conditions and permits determination of the enantiomeric excess. For dilute solutions, the degree to which a substance rotates plane-polarized light depends upon the number of molecules present in solution and their ability to interact with the light. This is in turn dependent upon the concentration of the solution, the path length of the cell and the wavelength of the light used for analysis. Commonly specific rotation is quoted for light at the wavelength of the D line of the emission spectrum of sodium (589.3 nm). The temperature of the sample and the nature of the solvent may also affect the value and these must also be stated when quoting specific rotation. The sign of the rotation must also be quoted. If clockwise it is +ve and if anticlockwise −ve.

$$\text{Specific rotation } [\alpha] = \pm \frac{100\alpha}{c \times l}$$

α = observed rotation of sample (°)
c = concentration of sample (g 100 mL^{-1})
l = path length of cell (dm)
t = temperature

Quoted as: $[\alpha]_D^t = \pm X$ (c = Y, solvent)

- Note the non-standard units for deriving [α] and the fact that, by convention, the figure is always quoted dimensionless.

The enantiomeric excess of a scalemic mixture can be deduced from the measured specific rotation.

$$\text{Enantiomeric excess} = \frac{100 \times \text{observed } [\alpha]_D}{[\alpha]_D \text{ pure enantiomer}} \%$$

E- and *Z*-alkenes

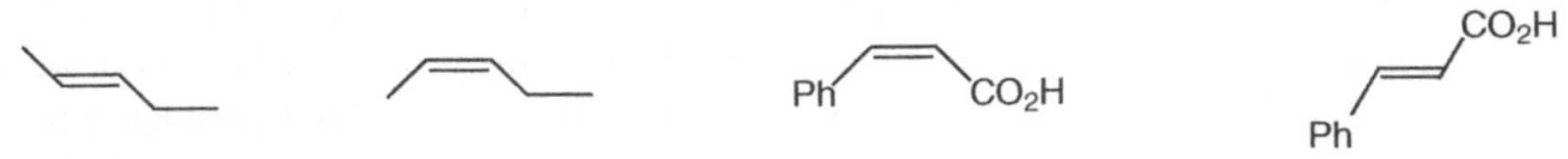

E-pent-2-ene

mp −140.2°C

bp 36.3°C

Z-pent-2-ene

mp −151.4°C

bp 36.9°C

Z-3-phenylpropenoic acid

mp 42°C

E-3-phenylpropenoic acid

mp 133°C

Compounds with more than one asymmetric centre exhibit *diastereoisomerism*

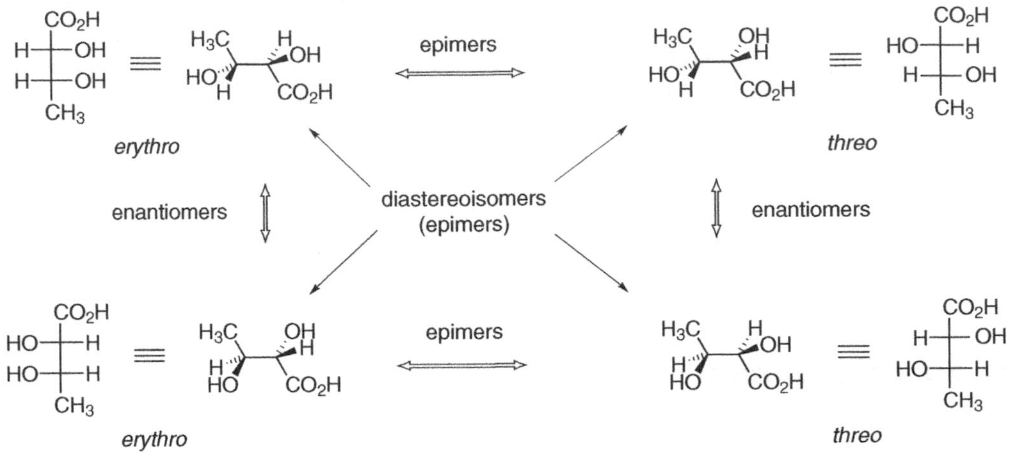

erythro

epimers

threo

enantiomers

diastereoisomers
(epimers)

enantiomers

erythro

epimers

threo

Molecular symmetry may reduce the number of diastereoisomers

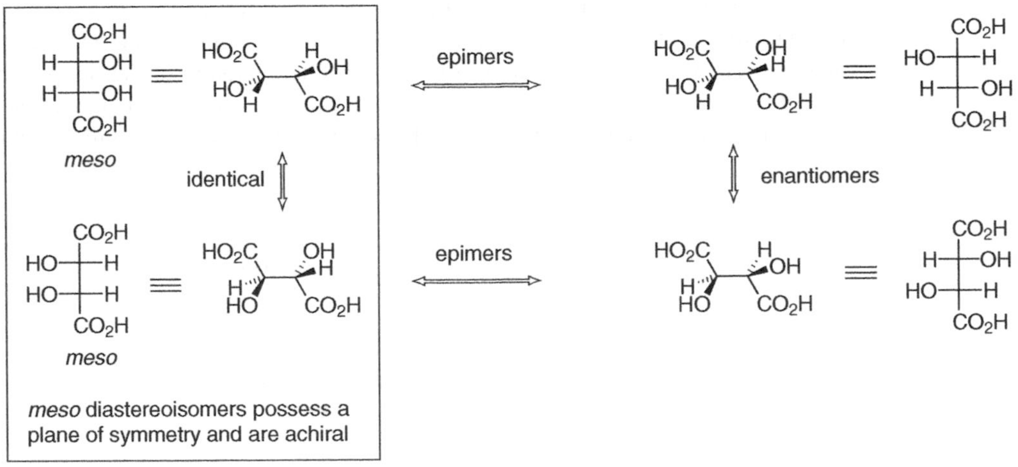

meso

epimers

enantiomers

identical

meso

epimers

meso

meso diastereoisomers possess a
plane of symmetry and are achiral

L-amino acids

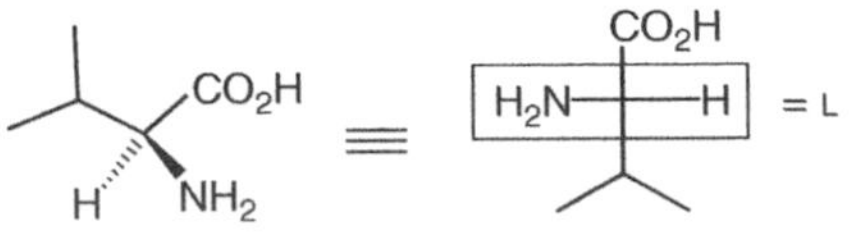

L-valine

H_2N——H = L

D-sugars

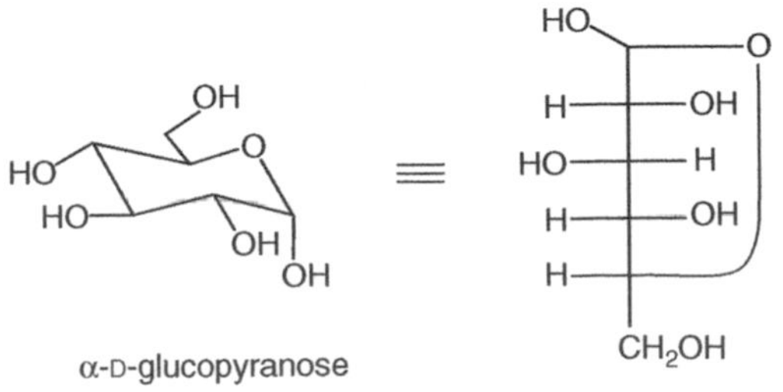

α-D-glucopyranose

3.6 Diastereoisomers (diastereomers)

Diastereoisomers are stereoisomers with a different *relative configuration* and are not related as mirror images. They have different chemical and physical properties.

The simplest type of diastereoisomerism occurs in alkenes, oximes and imines where interconversion of the double bond substituents is prevented by the energy barrier to rotation about the π-bond. The *E-isomer* (*entgegen*, German for *opposite*) has the highest priority substituents (see Section 3.2) on the double-bonded atoms pointing away from each other and the *Z-isomer* (*zusammen*, German for *together*) has the highest priority substituents on the same side. In the case of oximes and imines, the lone pair on the nitrogen is counted as the lowest priority substituent of that atom.

Molecules possessing more than one stereogenic centre also exhibit diastereoisomerism because inverting one or more (but not all) of the centres leads to structures which do not have a mirror image relationship with the original. Inversion of a single stereogenic centre gives an *epimer* of the original structure. Inversion of all stereogenic centres gives the *enantiomer*.

- A molecule possessing n stereogenic centres has a maximum of:

  2^n stereoisomers, 2^{n-1} pairs of enantiomers and n epimers.

- Molecular symmetry within the molecule may result in a reduction of the numbers of different isomers due to internal compensation.

3.7 Fischer projections

This convention proposed by Emile Fischer in 1891 attempts to depict molecular structure in a two-dimensional framework of vertical and horizontal bonds.

- The main carbon chain is drawn as a vertical line and bonds to all substituents are drawn as horizontal lines.
- All vertical lines represent bonds behind the plane of the page.
- All horizontal lines represent bonds in front of the plane of the page.

Rotating a Fischer projection by 90° or interchanging two substituents results in inversion of the absolute configuration of the stereogenic centre. The absolute configuration may be deduced by ranking the substituents according to the sequence rules (see Section 3.2), placing the lowest ranking substituent on a vertical axis and determining whether the priority of the remaining substituents descends clockwise (*R*-) or anticlockwise (*S*-).

3.8 Erytho-, threo- and meso-

Erythro- describes adjacent stereocentres possessing similar or identical substituents on the same side of the vertical axis of the Fischer projection.

Threo- describes adjacent stereocentres possessing similar or identical substituents on the opposite side of the vertical axis of the Fischer projection.

Meso- describes a molecule which is achiral, despite possessing stereogenic centres, due to the presence of an internal mirror plane. For a molecule possessing two stereocentres *meso*- corresponds to *erythro*-.

3.9 L and D nomenclature

In carbohydrates this formalism defines the configuration of the stereocentre furthest from the carbonyl carbon. It is also used for α-amino acids to define the configuration at the α-centre. When drawn as a Fischer projection with C-1 at the top, the *D-configuration* possesses the highest priority substituent of the stereocentre in question on the right-hand side of the projection; whereas the *L-isomer* has the highest priority group on the left-hand side.

Staggered (*anti*-periplanar) conformation of ethane

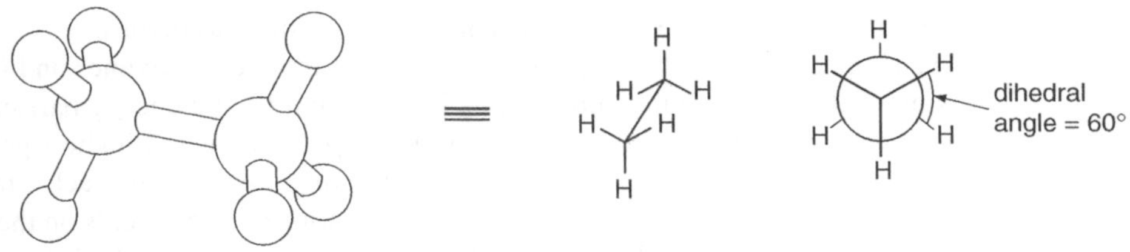

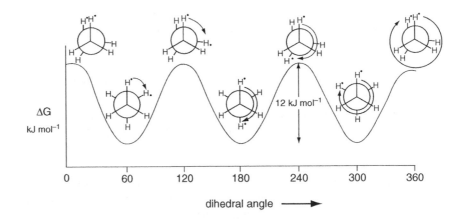

Ball and stick representation   **Saw horse** representation   **Newman** projection

dihedral angle = 60°

Eclipsed (*syn*-periplanar) conformation of ethane

Dihedral angle = 0°

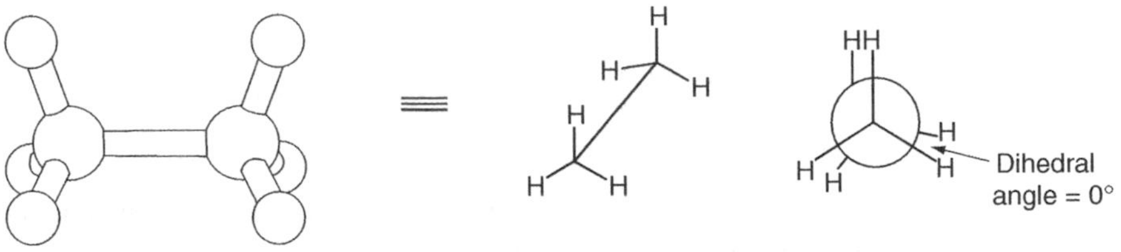

Relative energy of ethane conformers

ΔG kJ mol^{-1}

12 kJ mol^{-1}

| 0 | 60 | 120 | 180 | 240 | 300 | 360 |

dihedral angle ⟶

Relative energy of butane conformers

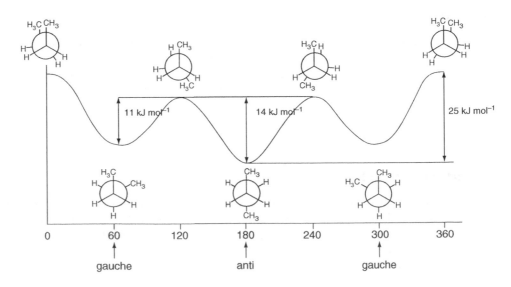

11 kJ mol^{-1} 14 kJ mol^{-1} 25 kJ mol^{-1}

| 0 | 60 | 120 | 180 | 240 | 300 | 360 |

gauche   anti   gauche

3.10   Conformation

The low energy barrier to rotation about single bonds means that most organic molecules can have an infinite number of three-dimensional arrangements of their constituent atoms known as *conformations* or *conformers*. Interconversion of *conformers* does not involve bond cleavage and care must be taken not to confuse conformation and configuration (see Sections 3.1 and 3.6).

Although an infinite number of conformations are accessible, certain of them possess energies which lie at local minima and these states will be more densely populated. Consideration of the energy changes which accompany rotations about single bonds is termed *conformational analysis*. O. Hassel and D. H. R. Barton were awarded the Nobel Prize in 1969 in recognition of their pioneering contributions in this area.

Two ways of illustrating conformational changes are the *sawhorse representation* and *Newman projection*.

3.11   Conformational analysis of acyclic molecules

Ethane

The most stable conformation is that in which the *dihedral angle* (ϕ) is 60°, often referred to as the *staggered conformation*. When the dihedral angle is 0° the conformation is said to be *eclipsed* and this corresponds to the energy maximum. The energy difference between these two states is $12\,kJ\,mol^{-1}$, reflecting repulsion between the electron pairs of opposed C–H bonds, and is the origin of the *torsional barrier*.

Butane

The same torsional strain analysis can be applied to rotamers about the C2–C3 bond of butane but the additional effect of van der Waals repulsion between the methyl groups results in a rise in energy of two of the staggered conformations (*gauche*) to $3.7\,kJ\,mol^{-1}$ higher than that of the most stable rotamer (*anti*) in which the methyl groups are disposed at a dihedral angle of 180°. To distinguish the different types of staggered and eclipsed conformations possible with butane, the relationship between the methyl groups is often described using *syn/anti* and *coplanar/periplanar* terminology. Thus the most stable conformation results when the two methyl groups are *anti-periplanar*.

3.12   Strain in cyclic molecules

In addition to strain associated with acyclic systems, cyclic molecules are subject to additional conformational constraints due to the loss of degrees of freedom associated with ring formation. Ring strain varies with the size of the ring.

small (3-, 4-membered)	*normal* (5-, 6-, 7-membered)
medium (8-, 9-, 10-, 11-membered)	*large* (>11-membered)

Torsional strain

Gauche interactions are obligate in all rings and three-, four- and five-membered rings suffer from eclipsing interactions.

Angle strain

This occurs in three- and four-membered rings where the internal bond angles are far smaller than the tetrahedral bond angle. All other rings can avoid angle strain by adoption of non-planar conformations.

Non-bonded interactions

These are long-distance interactions which arise from ring substituents impinging upon each other's van der Waals radii, effectively trying to occupy the same space. Specific examples include 1,3-diaxial interactions in cyclohexane chair conformers, the 1,4-(bowsprit) interaction in cyclohexane boat conformers and transannular interactions in medium and large rings.

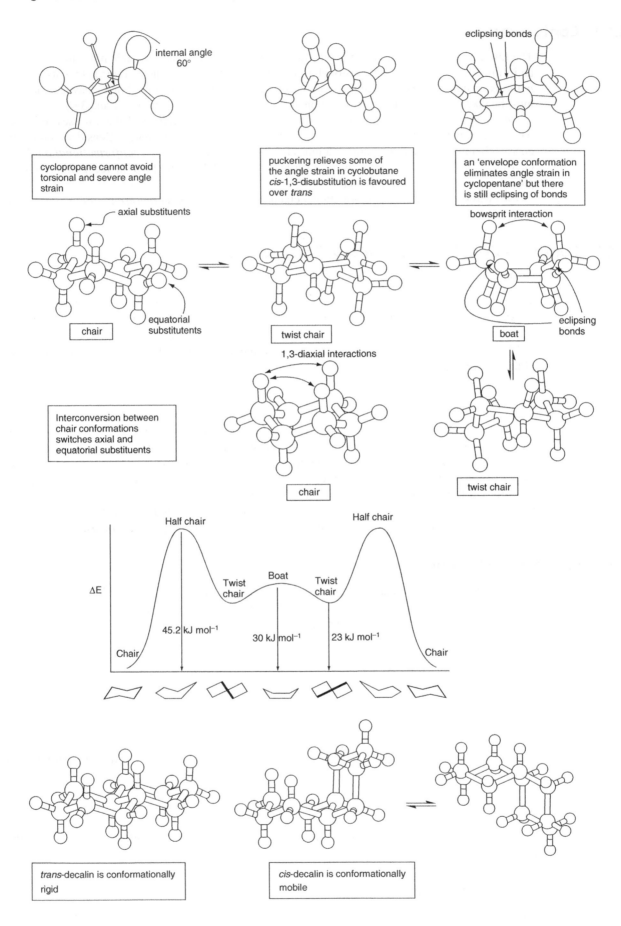

internal angle 60°

cyclopropane cannot avoid torsional and severe angle strain

puckering relieves some of the angle strain in cyclobutane *cis*-1,3-disubstitution is favoured over *trans*

eclipsing bonds

an 'envelope conformation eliminates angle strain in cyclopentane' but there is still eclipsing of bonds

axial substituents

chair

equatorial substitutents

twist chair

bowsprit interaction

boat

eclipsing bonds

Interconversion between chair conformations switches axial and equatorial substituents

1,3-diaxial interactions

chair

twist chair

ΔE

Half chair

Half chair

Twist chair

Boat

Twist chair

Chair

45.2 kJ mol⁻¹

30 kJ mol⁻¹

23 kJ mol⁻¹

Chair

trans-decalin is conformationally rigid

cis-decalin is conformationally mobile

3.13 Conformational analysis of monocyclic molecules

Cyclopropane

This molecule has no choice but to adopt a ***planar conformation***. Consequently it experiences high ***angle strain*** and ***torsional strain***.

Cyclobutane

Adopting a ***puckered conformation*** results in a reduction of angle strain and torsional strain. Consequently *cis*-1,3-disubstituted cyclobutane derivatives are more stable than their *trans*-counterparts.

Cyclopentane

Angle strain is eliminated completely by adoption of an ***envelope conformation***, but an ***eclipsing interaction*** remains.

Cyclohexane

The three important conformations of cyclohexane are ***chair***, ***twist chair*** and ***boat***.

The most stable conformations of cyclohexane are the two equivalent ***chair*** conformations. Their stability arises from the absence of eclipsing interactions. Substituents on the chair are either in the plane of the ring (***equatorial***) or above and below the plane (***axial***). Chair conformers readily interconvert via the ***twist chair***. This ***conformational inversion*** causes substituents which were originally equatorial to become axial and vice versa.

Substitution on the cyclohexane ring renders the two chair conformers energetically inequivalent because axial substituents experience unfavourable ***1,3-diaxial non-bonding interactions***. Consequently the lower energy chair conformer will have bulky substituents in equatorial environments and will be the dominant constituent of the dynamic equilibrium. *Note*: This is a ***conformational bias*** and both conformers are present. Reaction may proceed via either conformer as removal of either will be compensated by rapid re-equilibration.

The ***boat*** conformer of cyclohexane lies higher in energy than the previous conformations. The energetic disadvantage arises from two eclipsing interactions and a bowsprit interaction.

Larger rings

Increasing ring size leads to a greater number of populated conformations. Cycloheptane has four important conformations and cyclooctane has eleven. In medium rings, an important source of strain comes from ***transannular non-bonding interactions*** wherein substituents on opposite sides of the ring may approach within their van der Waals distances.

3.14 Conformational analysis of bicyclic molecules

Decalins (bicyclo[4.4.0]decanes) consist of two cyclohexane rings, fused either *trans* or *cis*, resulting in configurational isomers (see Section 3.1).

trans-Decalin

This isomer has a strongly preferred conformation in which both rings exist as chairs. Chair–chair interconversion of either ring is impossible as this would force the other ring to adopt a 1,2-*trans*-diaxial disposition which would require rupture of the four-carbon bridge.

cis-Decalin

This isomer has two preferred conformers in which both rings exist as chairs. Conformational inversion is easily accommodated by the structure but each conformer is subject to marked 1,3-diaxial non-bonding interactions.

4

Structure – Activity

Single-step reaction profiles

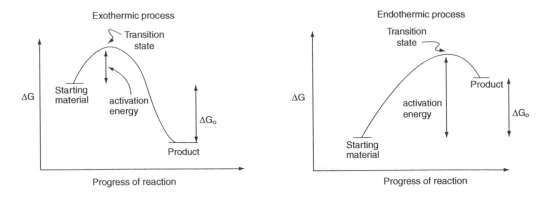

Two-step reaction profile

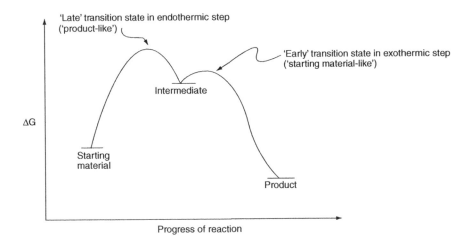

Kinetic versus thermodynamic control

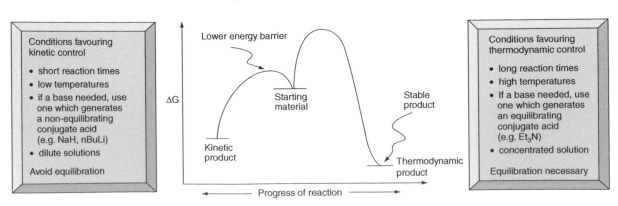

4.1 Reaction profiles

Single-step reactions

Chemical reactions may give out energy (***exothermic***) or absorb energy (***endothermic***). In either case progress from starting materials to products passes through a ***transition state*** which represents the point of highest energy in the reaction. With a knowledge of the energetics of a given reaction, it is possible to approximate the actual arrangement of the participating atoms at the transition state (see Section 4.2).

- Transition states are not isolable entities.

The difference in energy between the starting materials and the transition state is known as the ***activation energy*** and is the source of the ***kinetic barrier*** to reaction. The higher the activation energy, the more difficult it is to initiate the reaction.

Graphical representation of the change in overall energy of the system (***Gibbs free energy***, ΔG) during reaction is termed the ***reaction profile***.

Multi-step reactions

These result from sequential single step processes; the intervening products of each stage (***reactive intermediates***) exist at a local energy minimum and may be detected and sometimes even isolated.

- Unlike transition states, reactive intermediates have a finite lifetime.

4.2 Hammond postulate

In the *Journal of the American Chemical Society* in 1955, G. S. Hammond wrote:

'If two states, as for example, a transition state and an unstable intermediate, occur consecutively during a reaction process and have nearly the same energy content, their interconversion will involve only a small reorganization of molecular structure.'

The consequences of this postulate are as follows:

It is possible to approximate the arrangement of a transition state by considering the structure of the species (starting material, intermediate or product) at the closest local energy minimum.

An exothermic process has a transition state which is similar in nature to the starting material for that step (***early transition state***).

An endothermic process has a transition state which is similar in nature to the product of that step (***late transition state***).

- In an exothermic process, examination of the starting materials may permit qualitative assessment of the factors which will have greatest influence in the transition state; whereas for an endothermic process, examination of the products provides the same information.

4.3 Kinetic and thermodynamic control

Two factors control product distribution in competing reactions: ***relative rates*** (***kinetic control***) and ***relative stability of products*** (***thermodynamic control***). For kinetic control to be exerted, product equilibration must be prevented. Experimentally, this may be achieved by employing low temperatures and short reaction times.

Curtin–Hammett conditions

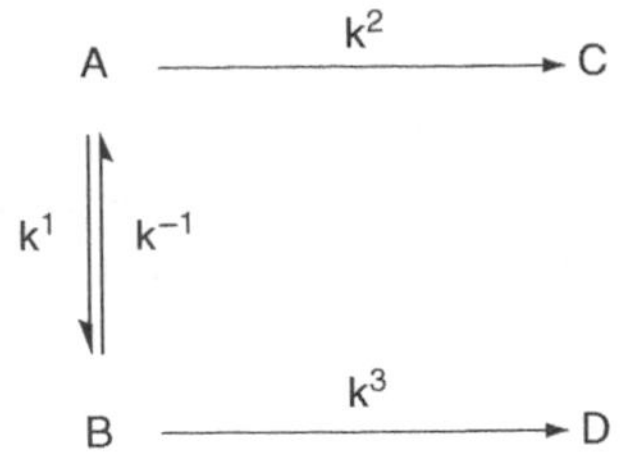

If $k^1, k^{-1} >>> k^2$ and k^3

$$\frac{[C]}{[D]} \propto \frac{k^2}{k^3}$$

Not dependent upon $\frac{[A]}{[B]}$

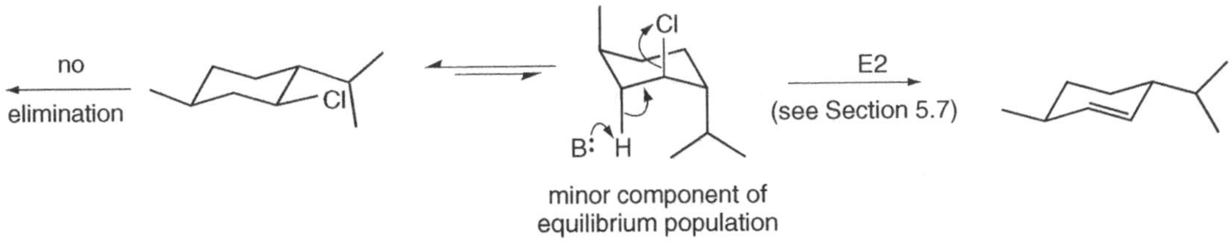

no elimination

E2 (see Section 5.7)

B: H

minor component of equilibrium population

Specific acid catalysis

No protonation in rate determining step, pH dependent

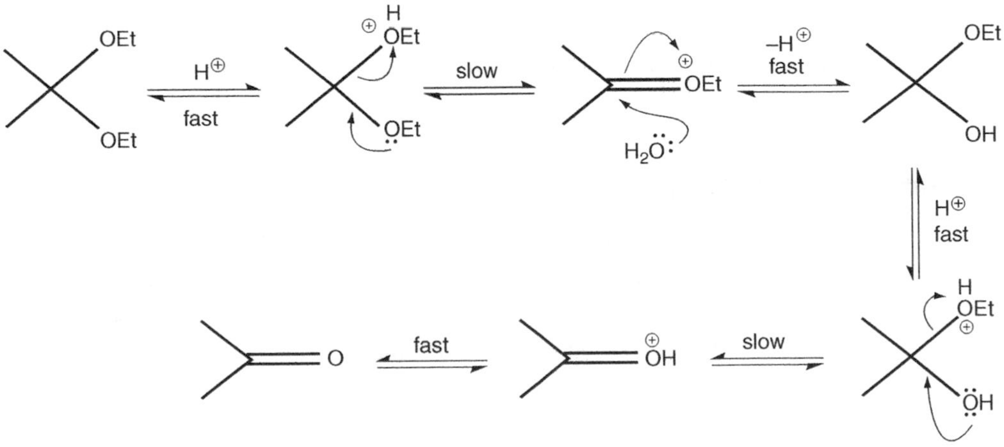

General acid catalysis

Protonation is rate determining step. Dependent on total concentration of all acid species

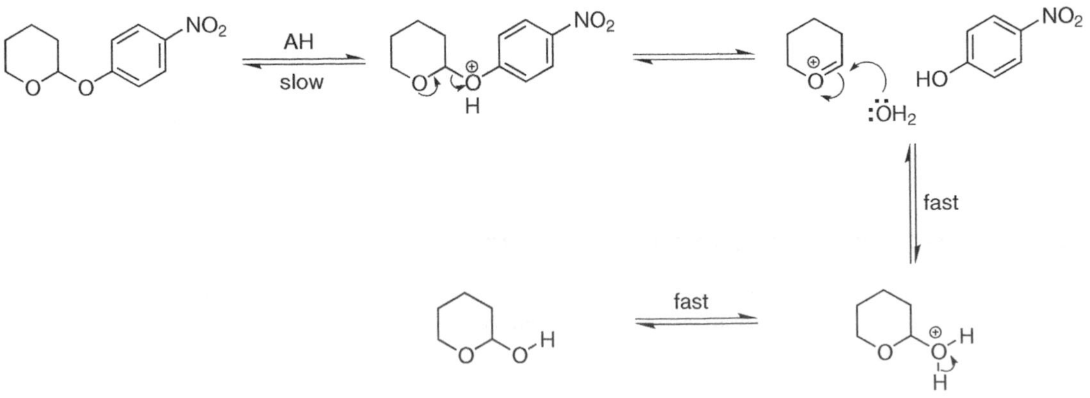

4.4 Curtin–Hammett principle

Molecules often exist as rapidly equilibrating mixtures of species, for example, conformers and tautomers. In situations where these initial states subsequently react by different pathways, the ratio of products is governed by the relative rates of the competing processes and is independent of the composition of the equilibrium mixture. Although originally applied to reaction conformers, this principle can be equally applied to other rapidly equilibrating systems.

- A reactive species existing as a very minor component of an equilibrating system can dominate the product distribution if it can be regenerated extremely rapidly.

4.5 Specific and general acid and base catalysis

These situations hold for reaction sequences in which protonation or deprotonation occurs as the initial step of a sequence.

Specific acid catalysis is dependent upon free protons, or hydroxonium ions (H_3O^+) if in aqueous solution. The initial protonation step is rapid compared to subsequent evolution of the protonated substrate which is rate limiting. Overall reaction rate is dependent upon the concentration of the protonated form of the substrate and increases with decreasing pH.

$$\text{For the reaction:} \quad A+H^+ \underset{k^{-1}}{\overset{k^1}{\rightleftharpoons}} AH^+ + B \xrightarrow[\text{r.d.s.}]{k^2} C$$

$$k^1 \gg k^2$$

$$\text{Rate} = k^2\,[AH^+]\,[B]$$
$$= K\,[H^+]\,[A]\,[B]$$

General acid catalysis results when protonation of the substrate is the rate-determining step. Thus, the overall reaction rate is dependent upon the concentrations of *all proton donors* present in the reaction mixture.

For the reaction: r.d.s.

$$A+HX \xrightarrow{k^1}$$
$$A+HY \xrightarrow{k^2}$$
$$A+HZ \xrightarrow{k^3} \left.\right\} AH^+ + B \xrightarrow{k^5} C$$
$$A+H^+ \xrightarrow{k^4}$$

$$k^5 \gg k^1, k^2, k^3, k^4$$

$$\text{Rate} = (k^1\,[HX] + k^2\,[HY] + k^3\,[HZ] + k^4\,[H^+])\,[A]$$

Specific base catalysis is dependent upon the concentration of *hydroxide ion* and reaction rate increases with increasing pH.

General base catalysis is dependent upon the concentrations of *all basic species* present in the reaction mixture.

- Maintaining constant pH but changing concentration of various proton sources (for example, by dilution) will alter the rate of general acid and base catalyzed reactions; whereas such changes will have no effect on reactions subject to specific acid or base catalysis.

4.4 Curtin–Hammett principle

4.5 Specific and general acid and base catalysis

5

Reaction Types

Unimolecular nucleophilic substitution

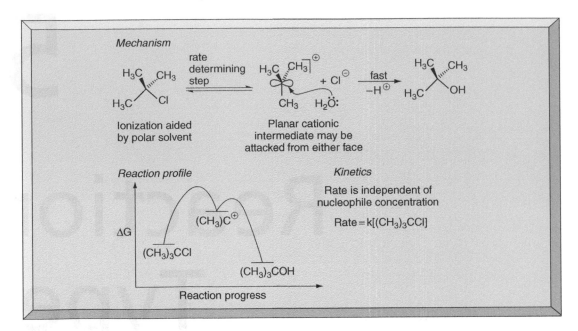

Bimolecular nucleophilic substitution

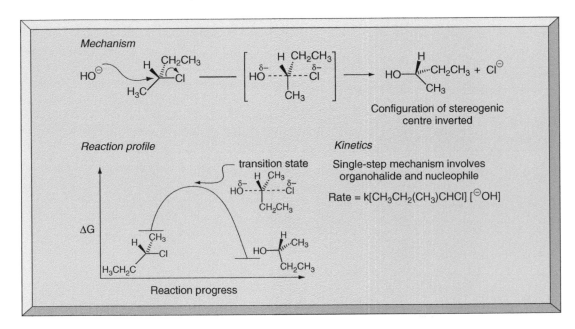

Less common nucleophilic substitution mechanisms

5.1 Aliphatic nucleophilic substitution

Replacement of one functional group by another at sp^3-hybridized carbon is referred to as *aliphatic substitution*. An incoming species which donates a pair of electrons is termed the *nucleophile* and a *leaving group* departing with a pair of electrons is termed the *nucleofuge*.

$$Nu: + R-X \rightarrow R-Nu + X:$$

If the incoming nucleophile is the solvent for reaction (for example H_2O), the reaction is known as *solvolysis*.

5.2 Unimolecular nucleophilic substitution (S$_N$1)

S$_N$1 reactions are *stepwise* processes in which the initial *rate-determining step* involves ionization of the alkyl substrate. The intermediate *trigonal planar carbocation* so formed reacts rapidly in a subsequent step with the nucleophile. As attack may occur on either face of the carbocation, enantiomerically pure substrates undergo racemization.

The rate of reaction is dependent only upon the concentration of the alkyl substrate and is independent of nucleophile concentration.

$$\text{Rate} = \frac{-d[RX]}{dt} = \frac{d[R^+]}{dt} = k[RX]$$

Formation of the carbocation is a highly endothermic process and, by reference to the Hammond postulate (see Section 4.2), factors influencing stability of the carbocation will operate on the transition state and will affect the overall rate of reaction.

5.3 Bimolecular nucleophilic substitution (S$_N$2)

S$_N$2 reactions are single-step processes in which attack by nucleophile and departure of nucleofuge occur simultaneously. Nucleophilic attack at the carbon centre occurs in a direction colinear with the bond being broken. The reaction proceeds via a penta-co-ordinate *trigonal bipyramidal transition state* to give a product in which the configuration of the carbon centre has been inverted (*Walden inversion*). The rate of reaction is dependent upon concentrations of both the alkyl substrate and the nucleophile.

$$\text{Rate} = \frac{-d[RX]}{dt} = \frac{d[RNu]}{dt} = k[RX][Nu]$$

5.4 Nucleophilic substitution with allylic transposition (S$_N$′)

This occurs when the substrate possesses a leaving group on a carbon attached to a double bond (*allylic substrate*). S$_N$1′ and S$_N$2′ reaction mechanisms are analogous to those of direct substitution except that nucleophilic attack occurs at the vinylic carbon γ- to the leaving group and the double bond is *transposed*. Direct substitution competes with allylic transposition and results in mixtures of products if the substituents at the α- and γ-positions are different. The nature of the substituents can influence the relative rates of direct and transposed substitution mechanisms.

5.5 Nucleophilic substitution with internal return (S$_N$i)

In rare cases, nucleophilic attack may occur on the same side of the substrate from which the nucleofuge departs resulting in *retention of configuration*. This stereoelectronically unfavourable process results from directed delivery of the nucleophile, often via a *tight ion pair*.

Unimolecular elimination

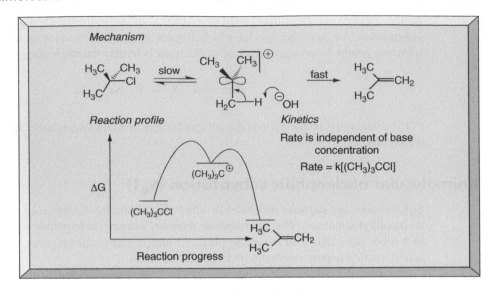

Bimolecular elimination

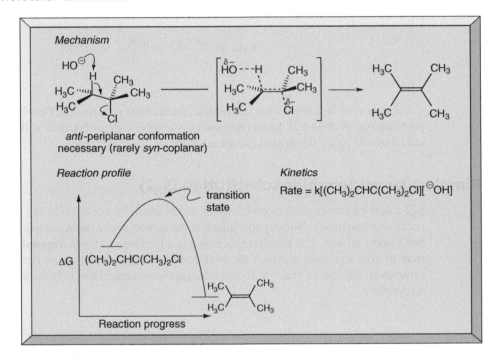

Unimolecular elimination from the conjugate base

Mechanism

$$PhSCH_2\text{--}CH_2CO_2Me \xrightarrow{\ominus OMe} PhS\text{--}CH_2\text{--}\overset{\ominus}{C}HCO_2Me \longrightarrow CH_2{=}CHCO_2Me + PhS^{\ominus}$$

Reaction favoured if carbanion is stabilized and with poor leaving groups

syn-Eliminations

Ester pyrolysis   *Selenoxide elimination*

5.6 Unimolecular elimination (E1)

Unimolecular elimination occurs when a carbocationic intermediate formed by loss of a nucleofuge undergoes rapid proton loss to form an alkene. The rate-determining ionization step is the same as in the S_N1 reaction and, indeed, E1 and S_N1 processes often compete. The reaction kinetics is first order and reaction rate depends only on substrate concentration.

$$Rate = \frac{-d[RX]}{dt} = \frac{d[R^+]}{dt} = k[RX]$$

As the carbocation has a finite lifetime, rotation around single bonds may lead to mixtures of configurational isomers in the product.

In situations where loss of different protons gives rise to regioisomers, the thermodynamically favoured, most substituted alkene predominates (***Saytzev product***).

5.7 Bimolecular elimination (E2)

By analogy with the S_N2 mechanism (see Section 5.3), bimolecular elimination proceeds via simultaneous deprotonation and departure of nucleofuge. At the transition state, both the C–H and C–nucleofuge bonds are partially broken and the alkene π-bond is partially formed. Base and substrate are involved in this single-step process and the reaction displays second order kinetics.

$$Rate = \frac{-d[RX]}{dt} = \frac{d[alkene]}{dt} = k[RX]\,[base]$$

Due to the geometry of the π-bond (see Section 2.3), the bonds which break during E2 elimination must lie in the same plane. The preferred conformation has the proton and nucleofuge in an *anti*-periplanar alignment (see Section 3.11), but if this is unattainable, reaction may proceed via a *syn*-coplanar arrangement. Hence, the relative configuration of the substrate controls the geometry of the alkene produced and the reaction is said to be ***stereospecific***.

In some cases, the availability of a suitably disposed β-proton can affect the direction of elimination and hence the position of the double bond. In addition, bulky bases, such as potassium *tert*-butoxide, will tend to remove the most accessible β-proton. Charged substituents (R_3N^+, R_2S^+) and fluorine, acidify β-protons due to inductive effects. This means that proton loss is more advanced than loss of nucleofuge at the transition state (E1cB-like, see Section 5.8) and the transition state possesses some carbanionic character. Factors which stabilize the carbanion will operate and often lead to the less substituted alkene (***Hofmann product***).

5.8 Unimolecular elimination from the conjugate base (E1cB)

Less commonly the proton β- to the nucleofuge is sufficiently acidified for complete deprotonation to occur before departure of the nucleofuge in the rate-determining step. Elimination rate is hence dependent upon the concentration of deprotonated substrate (***conjugate base***) and the reaction shows first order kinetics. This mechanism is most commonly observed in substrates possessing a nucleofuge β- to carbonyl or sulfone groups.

5.9 *syn*-Eliminations

Functional groups such as sulfoxides, selenoxides and tertiary amine oxides undergo unimolecular elimination by internal abstraction of a β-proton via a five-membered cyclic transition state. For such a mechanism to occur, a *syn*-β-proton is obligatory.

Reaction Types

S_N1 versus S_N2

S_N2	S_N1
Enhanced by	Enhanced by
Primary alkyl substrates	Stability of cation
	$1° < 2° < 3°$
Good nucleophile	Poor nucleophile
Good nucleofuge	Good nucleofuge
	Ionizing solvents

E1 versus E2

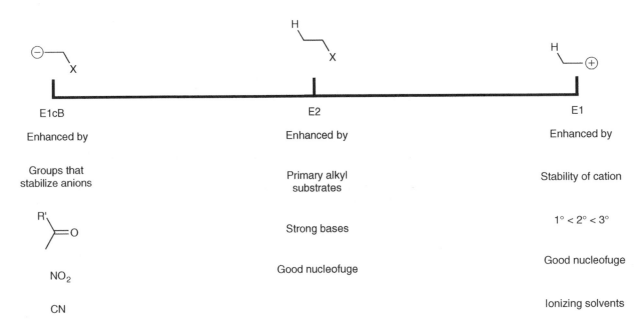

E1cB	E2	E1
Enhanced by	Enhanced by	Enhanced by
Groups that stabilize anions	Primary alkyl substrates	Stability of cation
	Strong bases	$1° < 2° < 3°$
NO_2	Good nucleofuge	Good nucleofuge
CN		Ionizing solvents
Ionizing solvents		

Bredt's rule

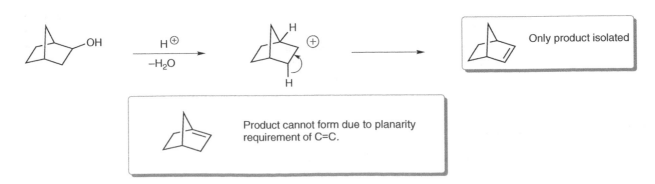

Only product isolated

Product cannot form due to planarity requirement of C=C.

5.10 S_N1 versus S_N2

There are no hard and fast rules governing the relative rates of S_N1 and S_N2 reactions and, in certain situations, both mechanisms may operate. The following are guidelines:

Alkyl substrate	S_N1	S_N2
$CH_3–X$	disfavoured	strongly favoured
$RCH_2–X$	disfavoured	favoured
$R_2CH–X$	intermediate	intermediate
$R_3C–X$	strongly favoured	disfavoured
$R_3CCH_2–X$	disfavoured	disfavoured
Good nucleophile	little effect	favoured
Good nucleofuge	favoured	favoured
Ionizing solvent	favoured	little effect
High temperature	strongly favoured	favoured

5.11 E1 versus E2

As E1 and E2 reactions are mechanistically related to S_N1 and S_N2 reactions respectively, it is not surprising that factors which affect nucleophilic substitutions also operate on elimination reactions. There are however, additional requirements imposed by the steric constraints of the E2 mechanism.

Alkyl substrate	E1	E2
$R_2CHCH_2–X$	disfavoured	favoured
$R_2CHRCH–X$	intermediate	intermediate
$R_2CHR_2C–X$	strongly favoured	disfavoured
Strong base	little effect	favoured
Good nucleofuge	favoured	favoured
Ionizing solvent	favoured	little effect
High temperature	strongly favoured	favoured

5.12 Substitution versus elimination

As there is such mechanistic similarity between substitution and elimination reactions, it is experimentally very difficult to achieve one or other pathway exclusively. The situation is complicated by the relative contribution of E1 versus S_N1 reactions and E2 versus S_N2 reactions to the overall process of substitution versus elimination.

- Elimination is favoured by high temperature and strong bases.
- Substitution is favoured by good nucleofuges.

5.13 Bredt's rule

Bredt's rule refers to the difficulty in introducing a double bond at the junction between two rings in bridged bicyclic compounds (***bridgehead position***). The structural constraints within such systems prevent coplanarity of the adjacent p-orbitals and consequently reduce π-overlap to the extent that double bond formation cannot occur (see Section 2.3). With larger ring systems, distortion of the framework occurs and bridgehead alkenes can be accommodated.

- It is not possible to introduce an alkene at a bridgehead of a bicyclic molecule if the ring containing the *E-* double bond is a seven-membered ring or smaller.

Electrophilic Addition

Symmetrical alkenes

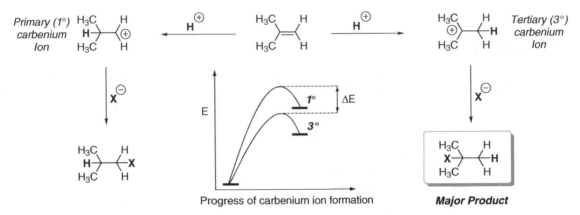

Intermediate carbenium ion

Unsymmetrical alkenes – Markownikoff's rule

Primary (1°) carbenium Ion

Tertiary (3°) carbenium Ion

Progress of carbenium ion formation

Major Product

Product is derived from the most stable carbenium ion.

Stereochemistry

HBr

Overall anti-addition of H and Br favoured.

Electrophilic halogenation

cyclic bromonium ion

Br₂

meso pair

Attack at either carbon of cyclic intermediate equally probable.

5.14 Electrophilic addition of hydrogen halides to alkenes

Symmetrical alkenes

The electron-rich π-bond (see Sections 2.3 and 2.4) will react with electrophilic reagents (see Section 1.5) such as HBr and HCl. The electrophile (in this case formally H^+) approaches the alkene from above or below the plane of the double bond. Transfer of the π-bond electron pair (see Section 2.4) occurs to form a C–H σ-bond, generating a carbenium ion (see Section 1.3) on the adjacent carbon. Subsequently, the carbenium ion is quenched by a nucleophile, which can be either another molecule of the acid or the halide ion.

- The addition of the electrophile is the rate determining step and dictates the course of the addition process, which is therefore termed *electrophilic addition* even though the second step of the process involves attack by a nucleophile.

Unsymmetrical alkenes (Markownikoff's rule)

With unsymmetrically substituted alkenes, two regioisomeric products may be formed, but generally one product predominates. An empirical rule used to predict the outcome was proposed by the Russian chemist Vladimir Markownikoff in 1905:

'In the addition of a hydrogen halide to an unsymmetrical alkene, the acidic hydrogen attaches to the less substituted carbon.'

The regiocontrol can be explained by considering the relative energies of the two possible transition states in the rate-determining addition of the electrophile H^+. This process is endothermic and has a late (product-like) transition state. Hammond's postulate (see Section 4.2) tells us that we can approximate the relative energy of these late transition states by considering the relative stability of the corresponding product carbenium ions. The predominant product is derived from the most substituted, most stable carbenium ion.

Stereochemistry

The predominant stereochemical course involves *anti*-addition of HX across the double bond. This result implies that either the carbenium ion is never fully formed, or that it is quenched more rapidly than rotation around the σ-bond, and that favoured approach by the nucleophile is *anti*-periplanar to the proton.

Electrophilic addition is effectively the reverse of elimination of HX from a haloalkane to form an alkene. This is an illustration of the *principle of microscopic reversibility* which recognizes that there is a single lowest-energy pathway for the interconversion of two states regardless of direction.

The pathway that is travelled in the forward direction of a reaction will also be followed in the reverse process.

5.15 Electrophilic halogenation of alkenes

Addition of bromine and chlorine to alkenes follows a mechanistically similar pathway to that of hydrogen halides. In this case, the electrophile is induced by polarization of the halogen bond on approaching the electron-rich π-bond. Formal addition of X^+ then occurs. In the case of bromine, this generates a cyclic *bromonium ion* which subsequently undergoes nucleophilic ring opening by backside attack of a second molecule of bromine to result in overall *anti*-dibromination of the double bond. In the case of chlorination the intermediate has more free carbenium ion character. Note that two molecules of halogen are involved in this addition, one supplying the electrophile and the other the nucleophilic partner.

Iodine adds reversibly to double bonds in an exclusively *anti*-manner, but it is not clear if the mechanism is polar or radical in nature. Fluorine adds violently and in an uncontrolled manner to give mixtures of products.

Free radical substitution of alkanes

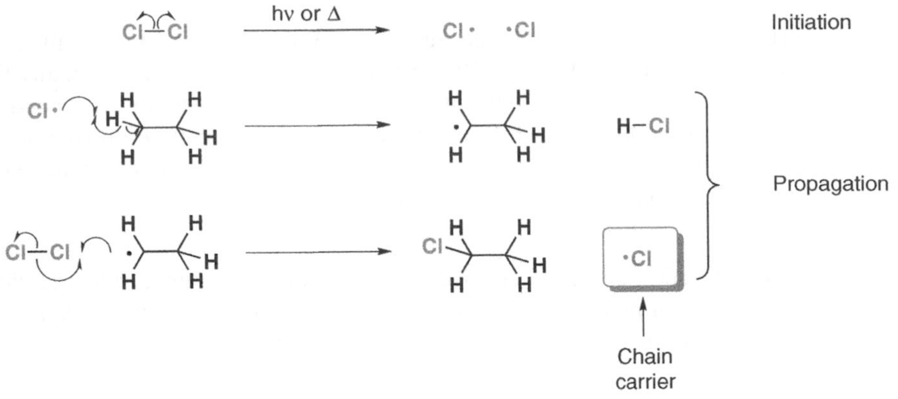

hν or Δ
Initiation
H–Cl
Propagation
•Cl
Chain carrier

Free radical allylic and benzylic substitution

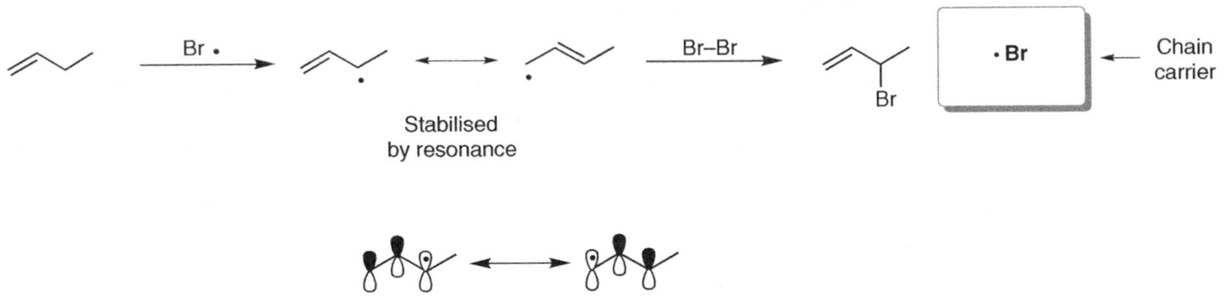

Br •
Br–Br
Br
• Br
Chain carrier
Stabilised by resonance

Free radical addition of hydrogen bromide to alkenes

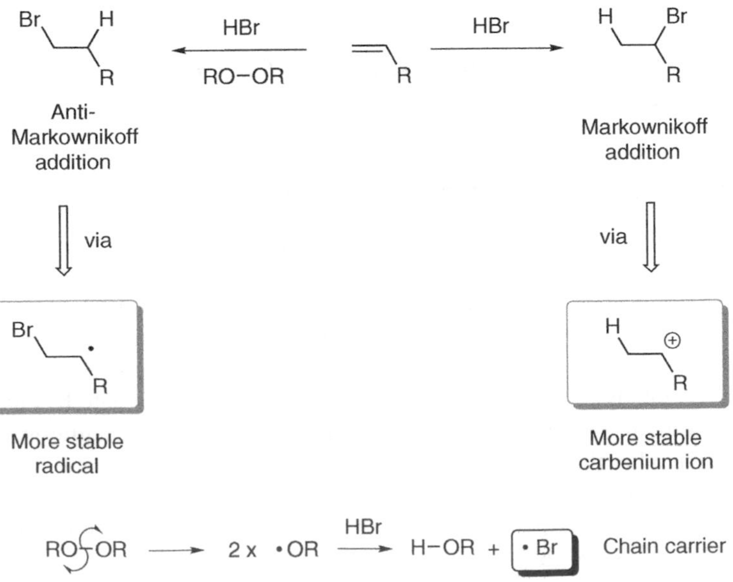

HBr
RO–OR
HBr
Anti-Markownikoff addition
Markownikoff addition
via
via
More stable radical
More stable carbenium ion
RO–OR
2 x •OR
HBr
H–OR + • Br
Chain carrier

5.16 Free radical substitution of alkanes

This constitutes an important method for the functionalization of otherwise unreactive hydrocarbons to form haloalkanes. The mechanism involves a *free radical chain reaction*.

The *initiation step* involves homolytic cleavage (see Section 2.7) of the halogen bond. This may be induced by thermal or photochemical means. The process is only synthetically useful for chlorine and bromine as the F–F bond is too strong to undergo ready cleavage (in any event, fluorine reacts uncontrollably with most organic substrates); whereas iodine radicals are too unreactive to participate in the subsequent chain reaction.

Propagation of the reaction is a two-step process. Initial *hydrogen atom abstraction* from the alkane results in the formation of an *alkyl radical*. This in turn reacts with another halogen molecule to produce the haloalkane and another halogen radical which re-enters the chain and continues the cycle.

Termination occurs when two radicals meet. This has the effect of removing reactive species from the system and causes the reaction to slow and ultimately stop. However, as the concentration of free radicals is always low, the likelihood of two radicals colliding is extremely small; although such reactions have very high rate constants. Under conditions where termination steps constitute minor pathways, a single initiation step results in many propagation cycles before the chain is terminated.

Free radical chlorination of alkanes results in a more or less statistical introduction of chlorine into the substrate reflecting the relative number of different types of hydrogen present. Bromine is more selective for the most substituted sites. Chlorine radicals are more reactive than bromine radicals and less selective in the product-determining hydrogen atom abstraction step. In contrast, hydrogen abstraction by bromine radicals is an endothermic process with a *late transition state* (see Section 4.2) and is more sensitive to the stability of the alkyl radicals formed. Alkyl radical stability mirrors that of carbocations ($3° > 2° > 1°$) and, in this case, leads to preferential formation of the most substituted bromoalkanes.

5.17 Free radical allylic and benzylic substitution

Resonance stabilization of allylic and benzylic radicals means that these are readily generated on hydrogen atom abstraction by a bromine or chlorine radical.

The halogen radicals may be generated by direct photo excitation of the halogen with UV light. Alternatively, under thermal conditions, a labile radical precursor is added to the reaction as the initiating agent. Compounds containing peroxy- and diazo-linkages are often used as these generate radicals at a synthetically useful rate.

Under either set of conditions, only low concentrations of radical species are present but the benzylic and allylic radicals are sufficiently long-lived to undergo bimolecular reaction with a molecule of the halogen. This results in overall *substitution* of halogen for an allylic or benzylic hydrogen.

Common sources of halogen are *N*-bromo- and *N*-chlorosuccinimide which possess weak nitrogen–halogen bonds.

5.18 Free radical addition of hydrogen bromide to alkenes

During the 1920s workers detected varying amounts of anti-Markownikoff products arising from the addition of HBr to alkenes. Subsequently it was deduced that low concentrations of peroxides present in the reaction mixtures were responsible for these observations and that intentional inclusion of peroxides led solely to anti-Markownikoff addition (*peroxide effect*).

Under these conditions, the peroxide-derived radical abstracts a hydrogen atom from HBr to generate low concentrations of bromine radicals which add to the double bond to form the more stable, more highly substituted carbon radical. This intermediate subsequently reacts with a further molecule of HBr, thus propagating the radical chain.

Electrophilic aromatic substitution

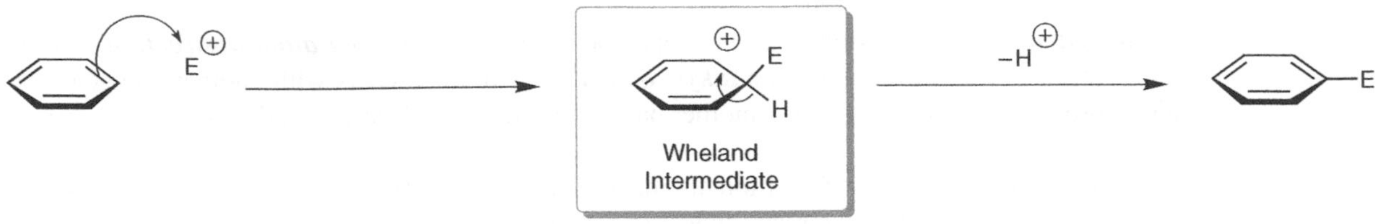

Wheland
Intermediate

Wheland Intermediate is stabilized by resonance

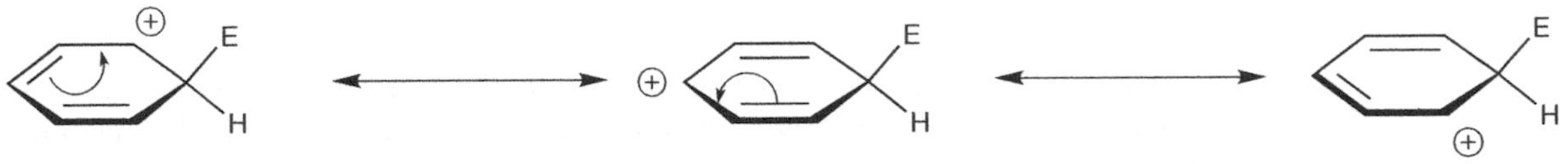

Monosubstituted benzene substrates

Activating, *ortho*, *para* directors

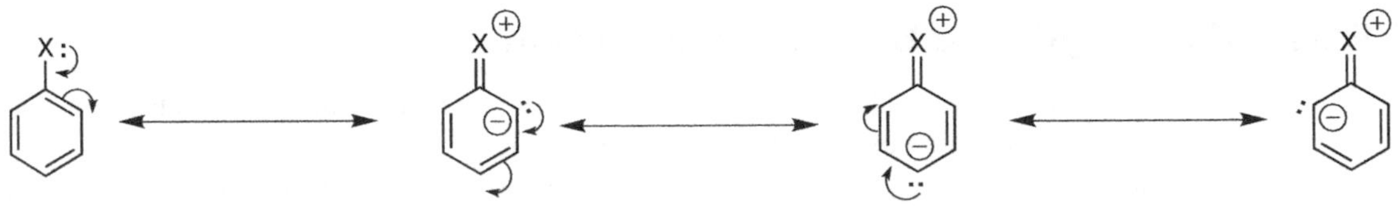

Overall

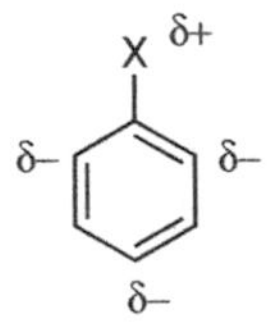

Higher electron density relative to benzene. Extra electron
density on carbons *ortho* and *para* to the substituent X.
Therefore electrophile is directed to these carbons.

Deactivating, *meta* directors

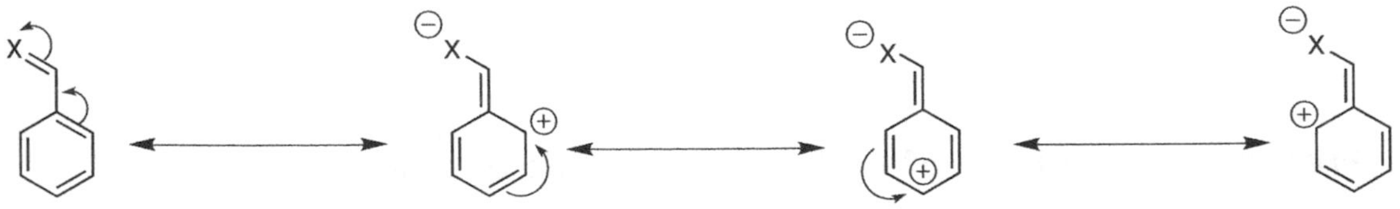

Overall

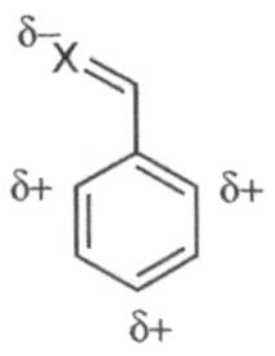

Lower electron density relative to benzene. Partial positive
charge on carbons *ortho* and *para* to the substituent X.
Therefore electrophile is directed to the *meta* position.

5.19 Electrophilic aromatic substitution

The aromatic stability of benzene (see Section 2.14) results in reduced reactivity of its π-system towards electrophiles compared with that of alkenes. Reaction with electrophiles results in overall *substitution* since this retains aromaticity in the product. This is in contrast to electrophilic addition reactions of alkenes (see Section 5.14).

5.20 Electrophilic aromatic substitution of benzene

Only very reactive species will react with benzene. These are commonly generated in the presence of a Lewis acid (see Section 1.4). Initial face-on approach results in the formation of a π-*complex* between the electrophile and the aromatic π-system. Subsequent electron-pair transfer forms a σ-bond and results in the generation of a reactive positively charged *Wheland intermediate*. Although lacking aromatic stability, the cyclohexadienyl cation is somewhat stabilized due to resonance (see Section 2.12). Loss of a proton from the Wheland intermediate re-establishes the aromatic system and leads to the substituted product.

5.21 Electrophilic substitution of monosubstituted benzene substrates

Electrophilic substitution of a monosubstituted benzene derivative can give rise to three regioisomeric products in which the new substituent is placed at C-2 (*ortho*), C-3 (*meta*), or C-4 (*para*) with regard to the original substituent. Due to symmetry, there are two equivalent *ortho* and *meta* positions. However, a statistical distribution of products is not observed and the actual proportion is dependent upon the electronic characteristics of the original substituent.

- *Activating groups* enhance the electron density of the aromatic π-system and thereby increase the rate of reaction with electrophiles.
- *Deactivating groups* reduce the electron density of the aromatic π-system and thereby decrease the rate of reaction with electrophiles.

Activating groups	Deactivating groups
alkyl	nitro
aryl	ester
hydroxyl	keto
ether	ammonium
thioether	sulfonium
amino	sulfonic acid
oxyanion	trihalomethyl
	halogen

- Substituents which are net electron donors (inductively or mesomerically) cause an increased rate of substitution compared to benzene and a predominance of 1,2- and 1,4-disubstituted products (*ortho/para activating groups*).
- Mesomerically withdrawing substituents cause a decreased rate of substitution compared to benzene and a predominance of 1,3-disubstituted products (*meta deactivating groups*).
- Halogens are strongly inductively withdrawing but mesomerically donating substituents which cause a decreased rate of substitution compared to benzene but a predominance of 1,2- and 1,4-disubstituted products (*ortho/para deactivating groups*).

Despite the fact that there are two *ortho*-positions but only one *para*-position, aromatic substrates possessing *ortho/para*-directing substituents tend to yield mainly 1,4-disubstituted products. This is a consequence of steric effects; generally the larger the original substituent, the greater the proportion of 1,4-disubstitution.

Mechanistic basis of directing effects

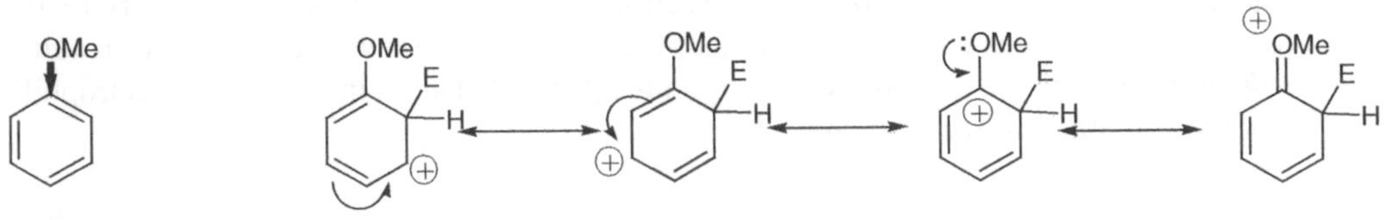

Mesomeric
Donating Group

Donation dominates, electron rich relative to benzene, electrophile attacks ortho or para.

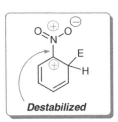

Mesomeric
Withdrawing Group

Withdrawal dominates, electron poor relative to benzene, electrophile attacks meta.

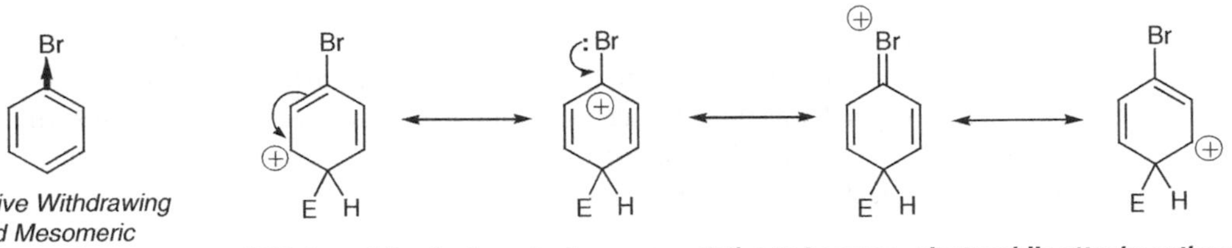

Inductive Withdrawing
and Mesomeric
Donating Group

Withdrawal dominates, electron poor relative to benzene, electrophile attacks ortho or para.

Partial rate factors

How reactive is *tert*-butylbenzene towards nitration compared with benzene?

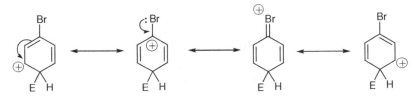

Experimentally derived
partial rate factors

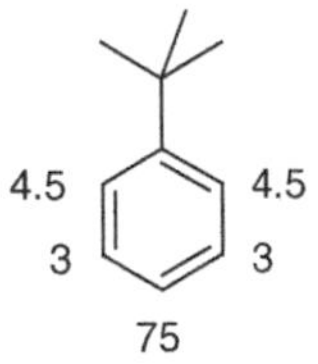

4.5   4.5
3   3
75

$$\frac{\textit{tert}\text{-Butylbenzene}}{\text{Benzene}} = \frac{2(4.5) + 2(3) + 75}{6(1)} = \frac{90}{6} = 15 = n$$

Predict the distribution among the various mono-nitration products of *tert*-butylbenzene.

f_{ortho} = $\frac{6}{2} \times \frac{x}{100} \times n$ Becomes: 4.5 = $\frac{6}{2} \times \frac{x}{100} \times 15$ ∴ x = 10%   *ortho*-nitro

f_{meta} = $\frac{6}{2} \times \frac{x}{100} \times n$ Becomes: 3 = $\frac{6}{2} \times \frac{x}{100} \times 15$ ∴ x = 6.7%   *meta*-nitro

f_{para} = $\frac{6}{2} \times \frac{x}{100} \times n$ Becomes: 75 = $\frac{6}{1} \times \frac{x}{100} \times 15$ ∴ x = 83.3%   *para*-nitro

5.22   Mechanistic basis of directing effects

The vast majority of electrophilic aromatic substitution reactions proceed under kinetic control in which the rate- and product-determining step is the initial generation of the Wheland intermediate. This is a highly endothermic process and Hammond's postulate (see Section 4.2) tells us that factors stabilizing the transition state will also operate in the Wheland intermediate. Hence, the more stable the intermediate, the faster its formation and the more that pathway contributes to product composition.

ortho/para-activating groups

The rate of substitution at all positions is enhanced compared to benzene due to the increased π-electron density of the aromatic ring. However, the Wheland intermediates resulting from *ortho*- or *para*-attack have the delocalized positive charge best placed to profit from maximum inductive or mesomeric stabilization.

meta-deactivating groups

The rate of substitution at all positions is slower than in benzene due to reduced π-electron density. The Wheland intermediates resulting from *ortho*- and *para*-attack are particularly destabilized; whereas the intermediate resulting from *meta*-attack is less destabilized.

ortho/para-deactivating groups

The inductive electron withdrawing effect of halogens reduces the π-electron density of the substrate and so rate of reaction at all positions is slower than with benzene. However, once formed, the Wheland intermediates resulting from *ortho*- or *para*-attack are stabilized by mesomeric electron donation of a lone pair from the halogen.

5.23   Partial rate factors

Partial rate factors (f) quantify the relative reactivity compared with benzene of various positions of substituted benzenes towards electrophilic substitution accounting for the fact that benzene has six equivalent positions.

For example, considering monosubstituted benzenes possessing two equivalent *ortho*, two equivalent *meta* and one *para* position.

- $f_{ortho} = 6/2$ (rate *ortho*/rate benzene)
- $f_{meta} = 6/2$ (rate *meta*/rate benzene)
- $f_{para} = 6/1$ (rate *para*/rate benzene)

Partial rate factors may be calculated from experimentally measured values of the overall rate of reaction of the substituted benzene compared to benzene itself and the product composition.

For a monosubstituted benzene which reacts n times faster than benzene to give a product mixture comprising x% *ortho*, y% *meta* and z% *para* products, partial rate factors are calculated as follows:

- $f_{ortho} = 6/2 \ (x/100 \times n)$
- $f_{meta} = 6/2 \ (y/100 \times n)$
- $f_{para} = 6/1 \ (z/100 \times n)$

Knowledge of partial rate factors therefore permits an appreciation of the relative nucleophilicities of various positions about a monosubstituted benzene compared to a single position on benzene.

Clearly, for polysubstituted benzenes, the multiplier in the calculations must be modified to account for the lower degree of symmetry of the substrate.

Nucleophilic aromatic substitution

F
:Nu
EWG
Nu
F
EWG
Meisenheimer complex
Nu
+
F
EWG

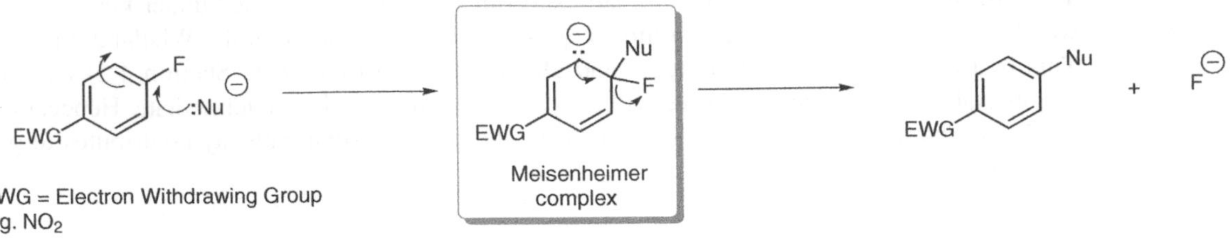

EWG = Electron Withdrawing Group
e.g. NO_2

Stabilization of the Meisenheimer complex

Nu
F
O
N
O
Nu
F
O
N
O
Nu
F
:O
N
O

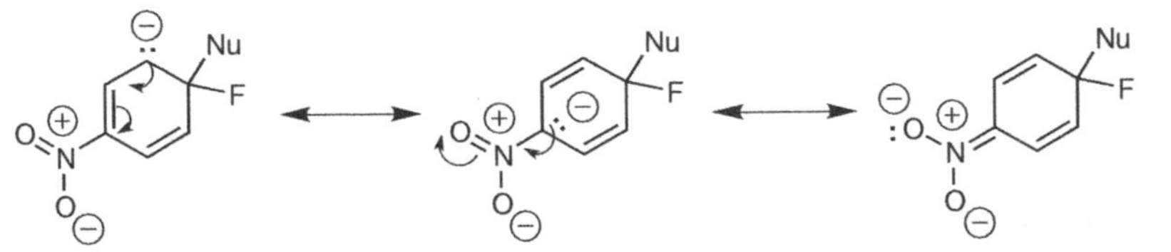

Mesomeric electron withdrawing groups *ortho* and *para* to the leaving group enhance reactivity by stabilizing the anionic intermediate most effectively.

Substitution of diazonium salts

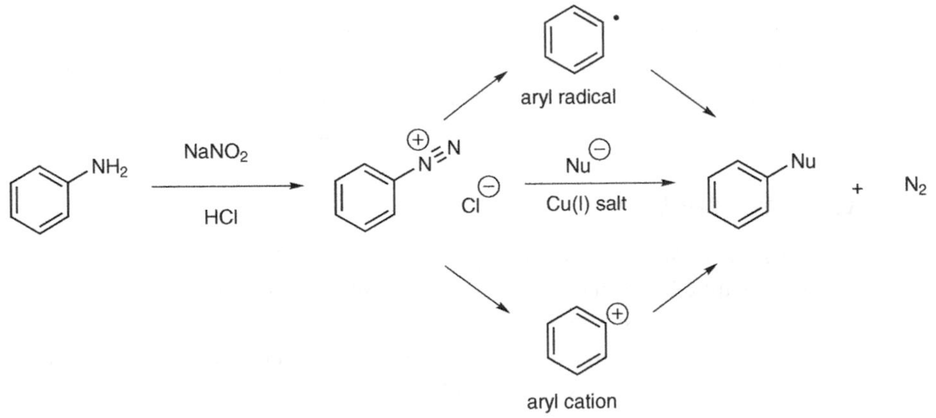

aryl radical
NH2
NaNO2
HCl
N≡N
Cl
Nu
Cu(I) salt
Nu
+
N2
aryl cation

Substitution via benzyne intermediates

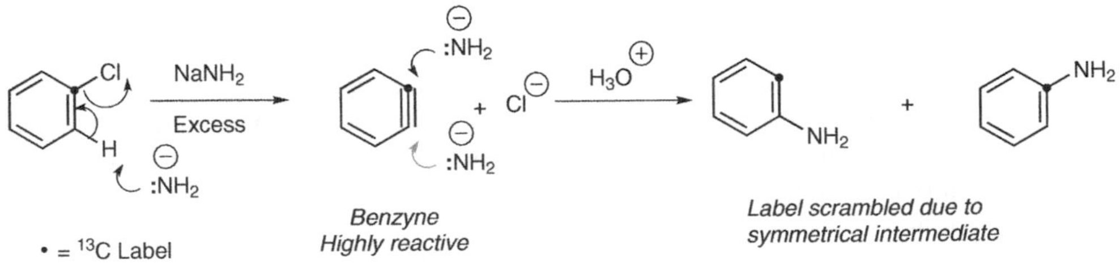

Cl
H
:NH2
NaNH2
Excess
:NH2
+
Cl
:NH2
H3O
NH2
+
NH2
• = 13C Label
Benzyne
Highly reactive
Label scrambled due to symmetrical intermediate

Electrophilic substitution of naphthalene

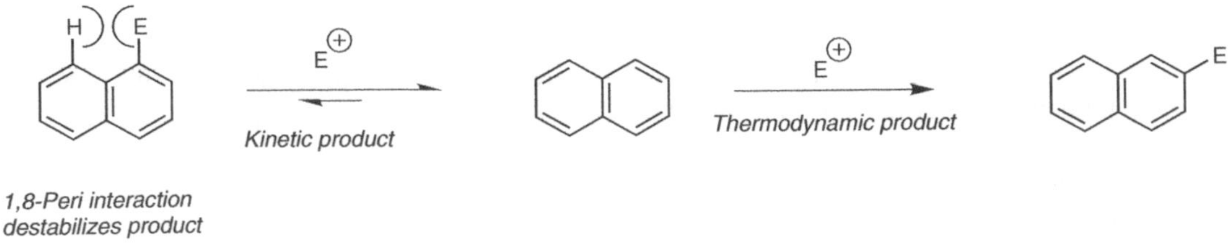

H
E
E
Kinetic product
E
Thermodynamic product
E
1,8-Peri interaction destabilizes product

5.24 Nucleophilic aromatic substitution (S$_N$Ar)

The high π-electron density of an aromatic ring results in predominant reactivity towards electrophiles. However, substituents such as halogens, the anions of which are good **nucleofuges** (see Section 5.1), may undergo displacement with strong nucleophiles. Reaction is favoured by electron withdrawing substituents on the ring in order to decrease π-electron density and facilitate approach of a nucleophile.

The mechanism proceeds via rate limiting nucleophilic attack at the site of substitution to give a negatively charged **Meisenheimer complex**. Subsequent loss of the nucleofuge regenerates the aromatic ring via an overall **addition–elimination** reaction.

The two-step mechanism is supported by the isolation of many Meisenheimer salts. Evidence for a rate-determining first step comes from the observation that fluoroaromatics undergo nucleophilic substitution much more rapidly than their iodo-counterparts, despite the fact that I$^-$ is a better nucleofuge than F$^-$. This is due to fluorine being more inductively electron withdrawing than iodine, reducing electron density on the aromatic ring and enhancing the rate of nucleophilic attack.

The high energy Meisenheimer intermediate is stabilized by resonance, resulting in higher electron density at the *ortho-* and *para*-positions. Hence, mesomerically withdrawing substituents (see Section 5.21) at positions *ortho-* or *para-* to the site of substitution are particularly effective in promoting S$_N$Ar reaction. This effect permits regioselective substitution with substrates possessing two identical leaving groups situated at different sites relative to the activating groups.

5.25 Substitution of diazonium salts (S$_N$1)

Diazonium salts readily undergo substitution of nitrogen with a variety of nucleophilic species. The process is stepwise and the mechanism of nitrogen loss may be either a purely heterolytic process, producing an **aryl cation**, or a homolytic process producing an **aryl radical**. The radical pathway is mediated by donation of a single electron from a metal species, for example Cu(I) → Cu(II).

Diazonium salts are prepared by reaction of aminobenzenes with nitrous acid at temperatures below 5°C as the salts are thermally unstable above this temperature, losing nitrogen and undergoing reaction with the water present to form phenols.

5.26 Substitution via benzyne intermediates

In this mechanism the nucleofuge is lost by 1,2-elimination to generate a highly reactive **benzyne** intermediate which traps nucleophilic species. This mechanism is restricted to systems in which a strong base is present (for example, NaNH$_2$, BuLi) and is commonly used with haloaromatic substrates.

Evidence for the elimination mechanism comes from the observation that isotopically labelled chlorobenzene undergoes scrambling of the label in the product, indicating the intermediacy of a symmetrical species. Furthermore 2,6-disubstituted halobenzenes will not undergo base-mediated reaction due to the requirement for an *ortho*-hydrogen.

5.27 Electrophilic substitution of naphthalene

Kinetically favoured substitution occurs at C-1 of the napthalene ring as the carbocation intermediate can be stabilized by resonance without disrupting aromaticity of the adjacent ring. However, the C-2 product is thermodynamically preferred as it does not suffer the 1-8 **peri-interaction** present in 1-substituted naphthalenes. Naphthalene is more reactive than benzene towards electrophilic substitution as aromaticity is retained in the non-reacting ring.

Electrophilic aromatic substitution of pyridine

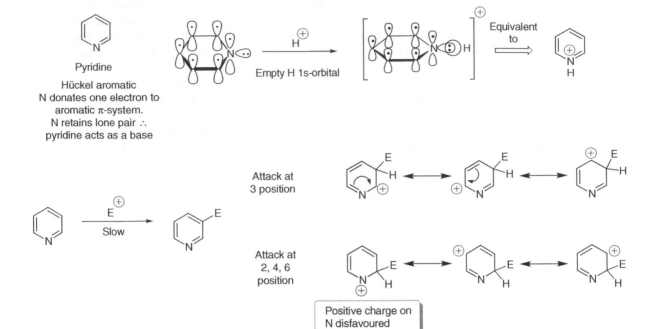

Pyridine

Hückel aromatic
N donates one electron to
aromatic π-system.
N retains lone pair ∴
pyridine acts as a base

Empty H 1s-orbital

Equivalent
to

Attack at
3 position

Attack at
2, 4, 6
position

Positive charge on
N disfavoured

Nucleophilic substitution of pyridine

**Chichibabin
reaction**

➤ *Nucleophilic attack at ortho and para positions as indicated.*

Electrophilic substitution of pyrrole

N donates two
electrons to the
π-system. Pyrrole
is non-basic.

Pyrrole is electron rich ∴ activated
towards substitution. All positions
have enhanced electron density ∴
pyrrole reacts with electrophiles at
C2 or C3.

Electrophilic substitution of indole, quinoline and isoquinoline

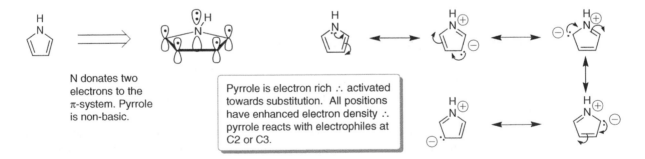

Indole: More reactive than benzene, reacts at C3.
Quinoline and Isoquinoline: Less reactive than
benzene, reaction is seen at C5 and C8.

5.28 Electrophilic aromatic substitution of pyridine

Pyridine (C_5H_5N) is the heteroaromatic analogue of benzene (C_6H_6) in which one of the CH groups has been replaced with nitrogen. The nitrogen lone pair is orthogonal to the aromatic π-system and has far reaching effects upon reactivity. Pyridine is a base and its *conjugate acid* has a $pK_a = 5.2$ (see Section 2.20).

Electrophiles react preferentially with the lone pair of the nitrogen to generate the *pyridinium ion* which, being positively charged, is unreactive towards electrophilic substitution. Neutral pyridine, which can react with electrophiles, is present only in a very low equilibrium concentration and so reaction rate is slow. Furthermore, pyridine is less reactive than benzene in this regard as the nitrogen polarizes the π-system, resulting in decreased electron density on the carbons. The electron withdrawing effect is strongest at positions 2, 4 and 6 and so electrophilic substitution to form 3-substituted products is least disfavoured to much the same degree that a nitro-substituent directs electrophilic substitution of benzene (see Section 5.19). Electron donating substituents at positions 2, 4 and 6 favour reaction.

The situation may be rationalized by comparing the Wheland intermediates resulting from attack at C-2 and C-4 with that arising from attack at C-3. Intermediates formed by the first two modes of reaction possess unfavourable resonance canonicals in which positive charge is located on nitrogen; whereas this is not the case with C-3 attack. This analysis is valid as the high-energy intermediates are good models for the transition states leading to them, in accordance with the Hammond postulate (see Section 4.2).

5.29 Nucleophilic aromatic substitution of pyridine

Pyridines are susceptible to nucleophilic attack at C-2 and C-4 as this leads to anionic intermediates which possess a favourable resonance canonical with the negative charge located on nitrogen. By analogy with nitrobenzene, 2- or 4-halopyridines will undergo preferential substitution of the halide compared to 3-halopyridines by a two step addition-elimination reaction (see Section 5.26).

Moreover, presence of a good nucleofuge at C-2 or C-4 is not necessary with certain nucleophiles (NH_2^-, alkyl and aryl anions) which will add, almost invariably at C-2, to form the dihydropyridine anion. This may be isolable, but usually undergoes subsequent oxidation to form the two-substituted pyridine. In the case of NH_2^- the reaction is known as the *Chichibabin reaction*.

5.30 Electrophilic aromatic substitution of pyrrole

The lone pair of the nitrogen in pyrrole is involved in the aromatic π-system (see Section 2.14) and resonance canonicals can be drawn in which negative charge resides on the carbons. Consequently, and in contrast to pyridine, pyrrole is reactive towards electrophilic substitution at either C-2 or C-3. By analogy with benzene, reaction involves addition of the electrophile, followed by regeneration of the aromatic ring by deprotonation. However, pyrrole is unstable towards oxidizing agents and therefore cannot be nitrated using a mixture of concentrated nitric and sulfuric acids.

5.31 Electrophilic aromatic substitution of indole

Indole shows reactivity comparable with that of pyrrole with the exception that electrophilic substitution occurs at C-3 as this can involve stabilization of the intermediate carbocation by mesomeric donation from the nitrogen without destruction of the homoaromatic ring.

5.32 Electrophilic aromatic substitution of quinoline and isoquinoline

Quinoline and isoquinoline are basic in the same way as pyridine and therefore aromatic substitution occurs on the homoaromatic ring as the heteroaromatic ring is deactivated due to protonation. Substitution is preferred at C-5 and C-8 as the intermediates can profit from resonance stabilization without disrupting the aromaticity of the heteroaromatic ring. *Note*: The numbering system for quinoline and isoquinoline does not follow standard IUPAC convention. In particular, numbering of the isoquinoline nucleus does not start at the heteroatom.

Carbenes

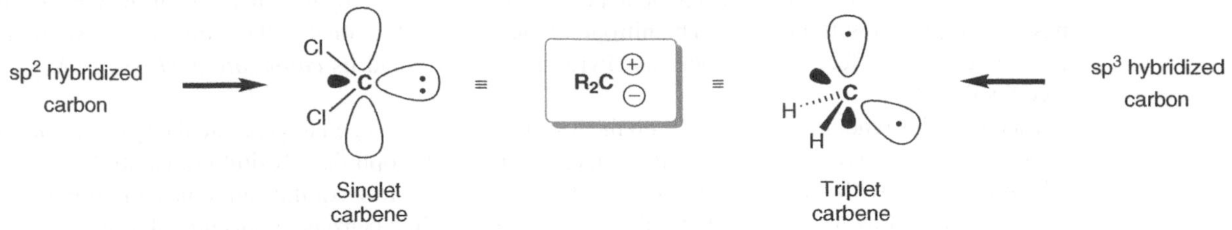

sp^2 hybridized carbon

Singlet carbene

Triplet carbene

sp^3 hybridized carbon

R_2C

1,1-Elimination

:OH$^{\ominus}$

Chloroform

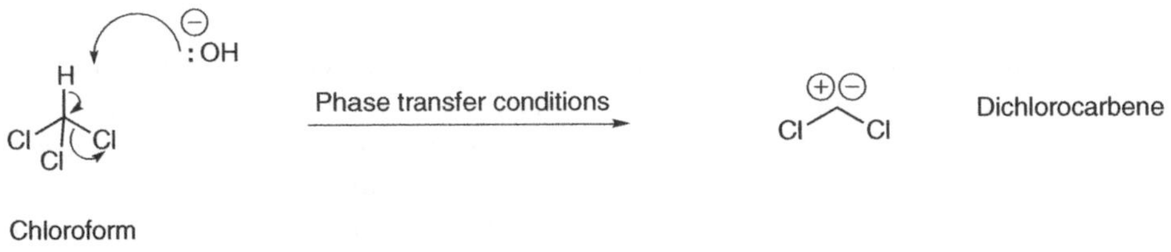

Phase transfer conditions

Dichlorocarbene

I—CH$_2$—I, Zn/Cu alloy

via   I—CH$_2$—ZnI $\equiv$ CH$_2$

Nitrenes

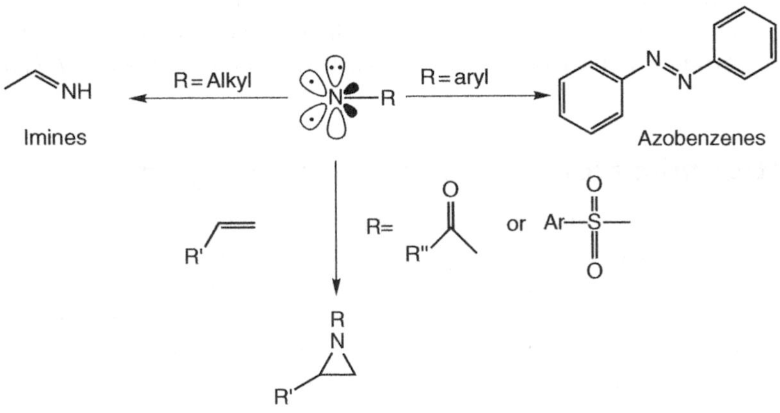

R = Alkyl

Imines

R = aryl

Azobenzenes

R =

or Ar—S—

Benzyne

−CO$_2$
−N$_2$

or

or

Nu

H$^{\oplus}$

[2+2] dimerization

(Z)   Diels–Alder cycloaddition

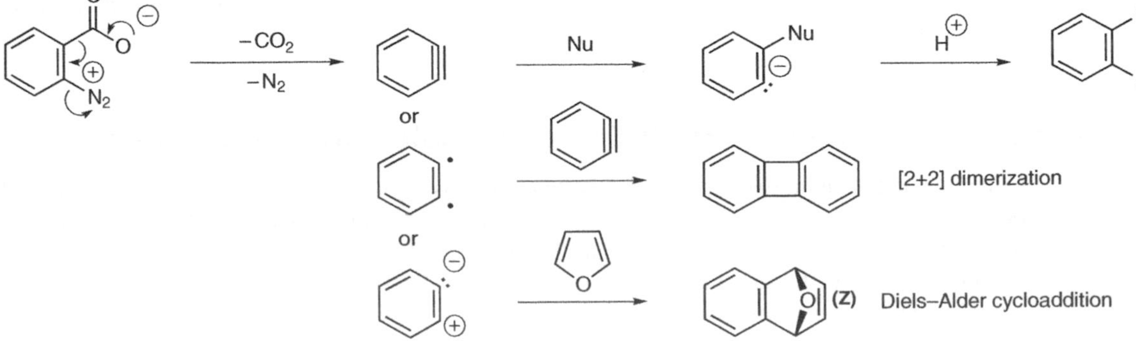

REACTIVE INTERMEDIATES

5.33 Carbenes

Carbenes are neutral, six-electron species which may exist in two electronic configurations. ***Singlet carbene*** is sp^2-hybridized, possessing an empty p-orbital; whereas ***triplet carbene*** is sp^3-hybridized and exists as a diradical. Dichlorocarbene exists in the singlet state; whereas carbene and alkyl-carbenes are usually generated in the triplet state. The electronic states lead to different reactivities, with singlet carbene undergoing concerted symmetry controlled insertions into double bonds (see Section 5.55) and triplet carbene stepwise stereorandom reaction. ***Note***: Whatever the electronic configuration, it is most simple to envisage the reactivity of a carbene by considering it formally to be 'R$_2$C$^{\pm}$'.

Carbenes may be generated by 1,1-elimination but, as this is less favourable than 1,2-elimination, substrates are limited to those lacking β-hydrogens. Phase transfer catalyzed base treatment of chloroform is a convenient way of generating dichlorocarbene. Photolytic extrusion of nitrogen from diazoalkanes or diazirines are preferred methods for preparing carbene or alkylcarbenes. Except for very rare examples, carbenes are too reactive to isolate and are generated and trapped *in situ*.

Carbenes insert into hydroxyl groups of alcohols or carboxylic acids, leading to ethers and esters respectively. Carbene insertion into alkenes occurs readily (see Section 5.55) and the organometallic species generated by reacting zinc–copper alloy with diiodomethane, formally I–CH$_2$–ZnI, is a convenient synthetic equivalent to carbene in the ***Simmons–Smith cyclopropanation***. Alternative carbene equivalents are phosphorus and sulfur ***ylids*** (see Section 5.54).

5.34 Nitrenes

Nitrenes are ***isoelectronic*** with carbenes, possessing a sextet of electrons in the outer valence shell, and are generally formed in the triplet state. Nitrenes may be generated by 1,1-elimination reactions or extrusion of nitrogen from azides.

Alkyl nitrenes rearrange to imines so rapidly that other reactions are excluded, although acyl and sulfonyl nitrenes will insert into alkenes to generate ***aziridines*** in an analogous reaction to carbenes. Aromatic nitrenes dimerize to give azobenzenes.

Nitrenes may be formally invoked in certain rearrangements in which there is migration to an electron-deficient nitrogen centre (see Section 5.59). However, involvement of a free nitrene does not necessarily occur and the stereospecific nature of the Beckmann rearrangement indicates a concerted process without generation of a free nitrene.

5.35 Benzyne

Benzynes are generated *in situ* by 1,2-elimination from halobenzenes. This process requires a strong base (for example NaNH$_2$, BuLi – see Section 5.26). Although benzynes are formally depicted as containing an alkyne, the linear geometry of such a unit (see Section 2.4) means that two sp-hybridized carbon atoms cannot be incorporated into a six-membered ring. Alternative ways of considering the structure are as a diradical or a charge separated species (see Section 2.12). The true nature of benzyne is presumably a hybrid of these possibilities resulting in a highly reactive, electrophilic species.

A non-basic method of generating benzyne involves diazotization of 2-aminobenzoic acid, leading to a diazonium salt which is not isolated, but allowed to undergo entropically favourable loss of CO$_2$ and N$_2$.

Benzyne will react with nucleophilic species, resulting in overall nucleophilic substitution either at the original site of substitution or at the position adjacent to it (***cine substitution***) (see Section 5.26). Benzynes may be trapped with dienes such as furan leading to products arising formally from Diels–Alder cycloaddition.

Reaction at the carbonyl group

Formaldehyde

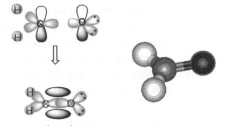

π bond

Significant resonance canonical

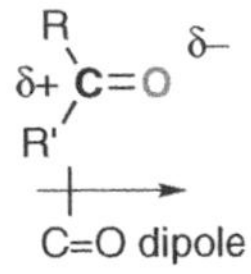

C=O dipole

As a consequence of the strong polarization, the carbonyl group itself exhibits the general reactivity shown

Nucleophilic reagents (electron rich or δ-) react at **electrophilic** carbon.

Electrophilic reagents (e.g. H⁺) react at *nucleophilic* oxygen

Carbonyl polarization in RCOX

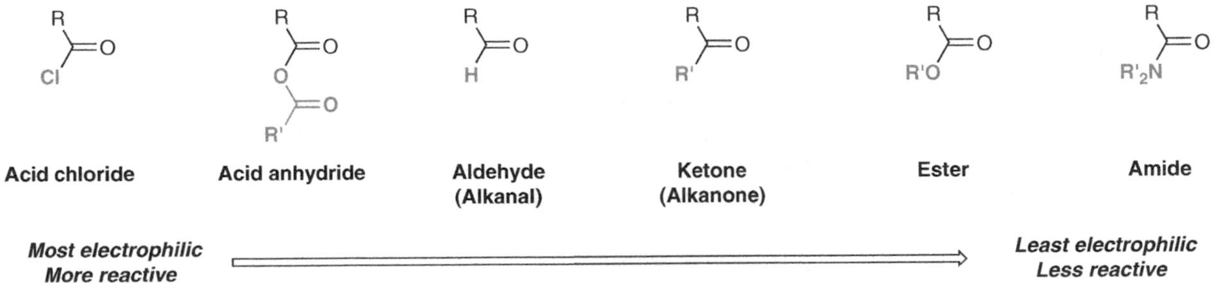

Acid chloride **Acid anhydride** **Aldehyde (Alkanal)** **Ketone (Alkanone)** **Ester** **Amide**

Most electrophilic
More reactive

Least electrophilic
Less reactive

1,2-Addition to the carbonyl group

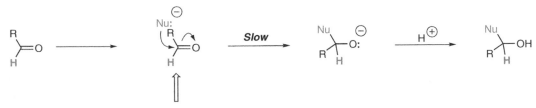

Slow

Nucleophile approaches at 109°
The Bürgi–Dunitz angle

Maximum overlap with π*

Repulsion from filled π-orbital forces attack at an obtuse angle

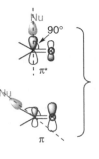

Combined effect – nucleophile attacks at **109°**

Bürgi–Dunitz angle

5.36 Reaction at the carbonyl group

The carbon–oxygen double bond of a carbonyl group consists of a σ-bond and a π-bond in much the same way as in alkenes (see Section 2.4). The non-bonding *lone pair* electrons on the oxygen and the substituents on the carbon lie in a plane orthogonal to the π-orbital and the oxygen behaves as if it were sp^2-hybridized.

The greater *electronegativity* (see Section 2.11) of oxygen compared to carbon (oxygen=3.5, carbon=2.5 relative to hydrogen=2.1) results in inductive polarization of the bond towards oxygen. A far greater polarizing effect is due to *resonance* (see Section 2.12) in which a significant contributing *extreme canonical form* has a negative charge located on the oxygen. Overall, these effects result in a substantial dipole moment, with the carbonyl carbon being *electrophilic* and the carbonyl oxygen *nucleophilic* (see Section 5.37).

5.37 Carbonyl polarization in RCOX

The nature of the carbonyl substituent X modifies the polarization of the carbonyl group, and hence the electrophilicity of the carbonyl carbon, through a balance of inductive electron withdrawal and mesomeric electron donation.

X	*Most electron withdrawing*		*Least electron withdrawing*			
Functional	halide	alkanoate	hydrogen	alkyl	alkoxy	amino
group	acid	acid	aldehyde	ketone	ester	amide
	halide	anhydride				
	Most electrophilic		*Least electrophilic*			

5.38 1,2-Addition to the carbonyl group

Depending upon conditions, the carbonyl group reacts initially either by oxygen attacking an electrophile or by a nucleophile attacking the carbon. This is followed by reaction with a nucleophile or electrophile to generate a tetrahedral species. Both steps are reversible in either pathway and the position of equilibrium is dependent upon the nature of the carbonyl substrate and the reagent system. Regardless of the order of the additions, nucleophilic attack is the *rate-determining step*. For this reason, the reactivity of the different carbonyl derivatives follows the relative electrophilicity of the carbonyl carbon. The order of reactivity of carbonyl derivatives towards nucleophilic attack is as follows:

acid halide	>	acid anhydride	>>	aldehyde	>	ketone	>	ester	>	amide
RCOCl		RCOOCOR		RCHO		RCOR′		RCOOR′		RCONR$_2$′

With certain combinations of nucleophile and carbonyl substituent, the adduct may be stable. The equilibrium is displaced to the right and the overall result is *1,2-addition* across the carbonyl group. This is a common mode of reaction for aldehydes (X=H) and ketones (X=alkyl, aryl) (see the following section).

The nucleophile approaches the rear of the carbonyl carbon in the π-plane at an angle of 109° (the *Bürgi–Dunitz* angle) above or below the plane described by the substituents on the carbonyl group.

Cyanohydrin formation

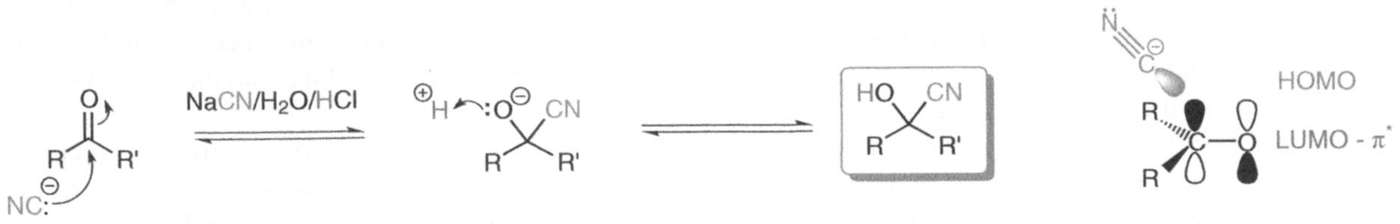

Grignard reagents

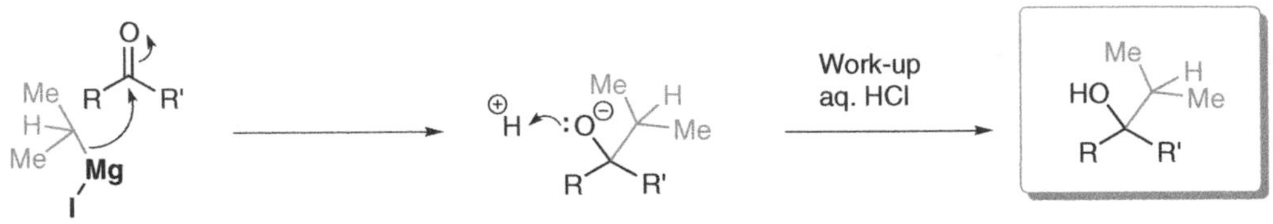

Conjugate addition

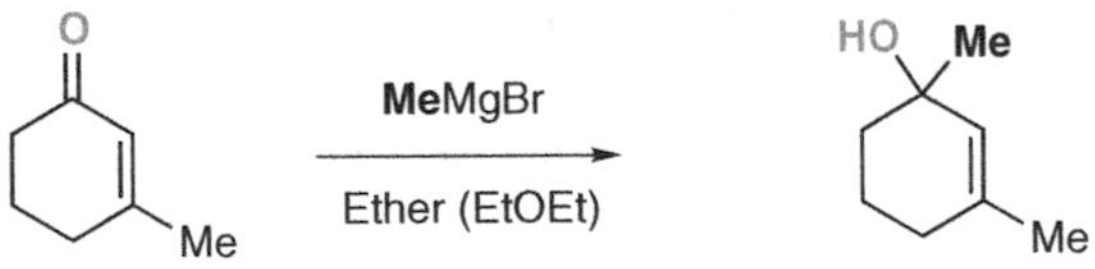

1,2-Addition favoured

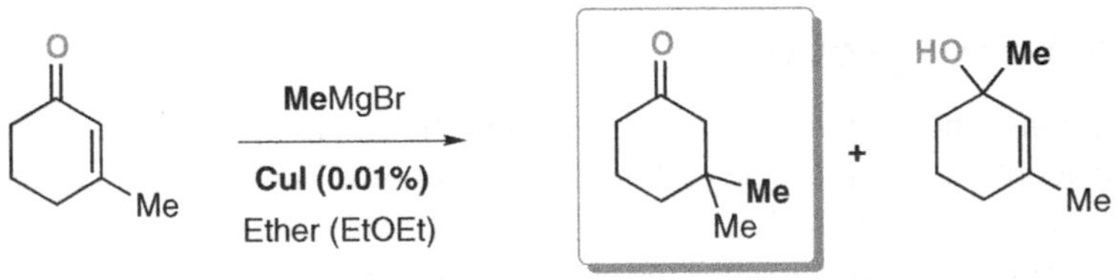

1,4-Addition favoured

Product Ratio:- 93 : 7

Organo copper species formed *in situ*

2 x MeLi + CuI ⟶ **(Me)₂CuLi**   Lithium dimethylcuprate

Mechanism

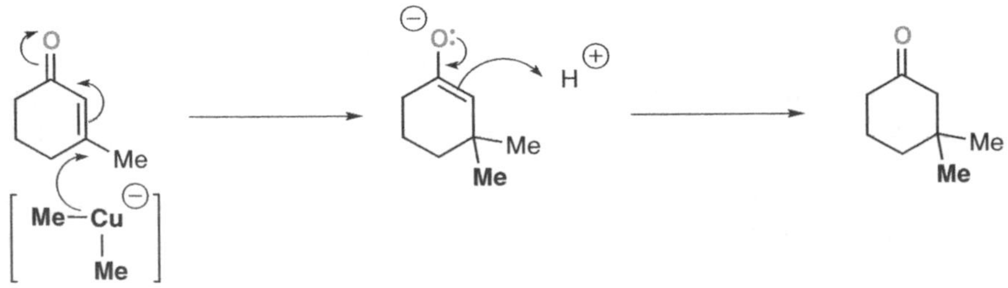

5.39 Reaction of aldehydes and ketones with carbon nucleophiles

Cyanohydrin formation

Under basic conditions, cyanide anion will undergo rate determining reversible nucleophilic attack on aliphatic aldehydes and ketones. A proton donor, such as water or HCN, causes the equilibrium to be displaced towards the *cyanohydrin*. Aromatic carbonyl compounds may form cyanohydrins but other reactions may compete.

Grignard reagents

These commonly used organometallic reagents (RMgX) act as a source of carbanions and add across the carbonyl group to generate the halomagnesium alkoxide. Mild acid hydrolysis of the intermediate gives an alcohol product.

- Methanal reacts with Grignard reagents to give primary alcohols.
- Other aldehydes react with Grignard reagents to give secondary alcohols.
- Ketones react with Grignard reagents to give tertiary alcohols.

As Grignard reagents are basic as well as nucleophilic, α-deprotonation to generate an enolate may compete with the addition. In this case, protonation of the enolate on work-up regenerates the starting carbonyl compound. Increasing steric hindrance around the carbonyl group and increased steric bulk of the Grignard reagent both disfavour nucleophilic addition and deprotonation can become the dominant process.

α-Branched Grignard reagents possessing β-hydrogens may undergo yet another side reaction. Transfer of a hydride from the β-position of the Grignard reagent to the carbonyl group via a cyclic six-membered transition state may occur, resulting in overall reduction in the alcohol after work-up.

Organolithium reagents

In general organolithium reagents are both stronger bases and stronger nucleophiles than Grignard reagents. Usually competing α-deprotonation is more troublesome than with Grignard reagents but, in the case of hindered ketones lacking α-hydrogens, organolithium reagents give better yields of carbonyl addition products.

5.40 α,β-Unsaturated carbonyl compounds as special substrates

In the case of α,β-unsaturated carbonyl compounds, the electropositive nature of the carbonyl carbon is relayed to the β-carbon. This can be illustrated by drawing the extreme resonance canonical forms (see Section 2.12). Therefore α,β-unsaturated carbonyl compounds are *ambident electrophiles* and nucleophiles may attack either the carbonyl carbon or the β-carbon. This latter process is often referred to as *conjugate addition* or *1,4-addition*.

Conjugate addition is referred to as 1,4-addition because the rate determining step involves nucleophilic attack on carbon at position 4 and electrophilic attack on oxygen at position 1. Subsequent tautomerism to the keto form on work-up masks the true nature of the initial addition process.

Conjugate additions of stabilized carbanions, such as those derived from 1,3-dicarbonyl compounds, constitute a class of reactions commonly referred to as *Michael additions*.

Steric hindrance around the carbonyl group favours conjugate addition; whereas substituents on the β-carbon favour attack at the carbonyl. A complex combination of factors governs the regioselectivity of attack by the nucleophile but a useful rule of thumb is that the *softer* the nucleophile, the greater the proportion of conjugate addition. Generally organolithium reagents add to the carbonyl; whereas the situation for Grignard reagents is less predictable. A particular type of organometallic reagent used specifically for conjugate addition is the *dialkyllithiocuprate R₂CuLi*, which is a *soft* nucleophile.

Hydration

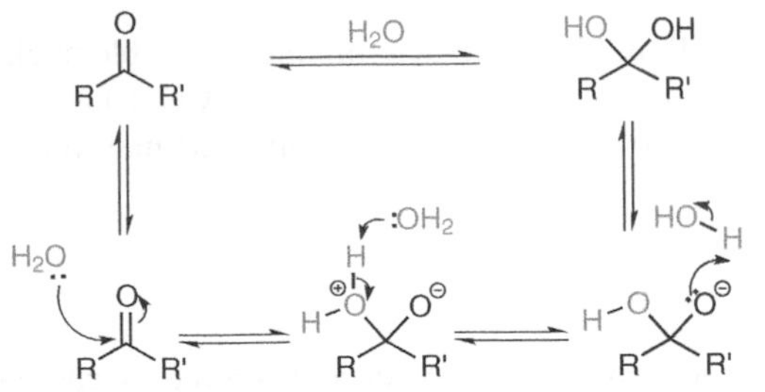

R	R'	K
CH_3	CH_3	0.001
CH_3	H	1
H	H	2280
CCl_3	H	2000
CF_3	CF_3	1200000

Acetal formation

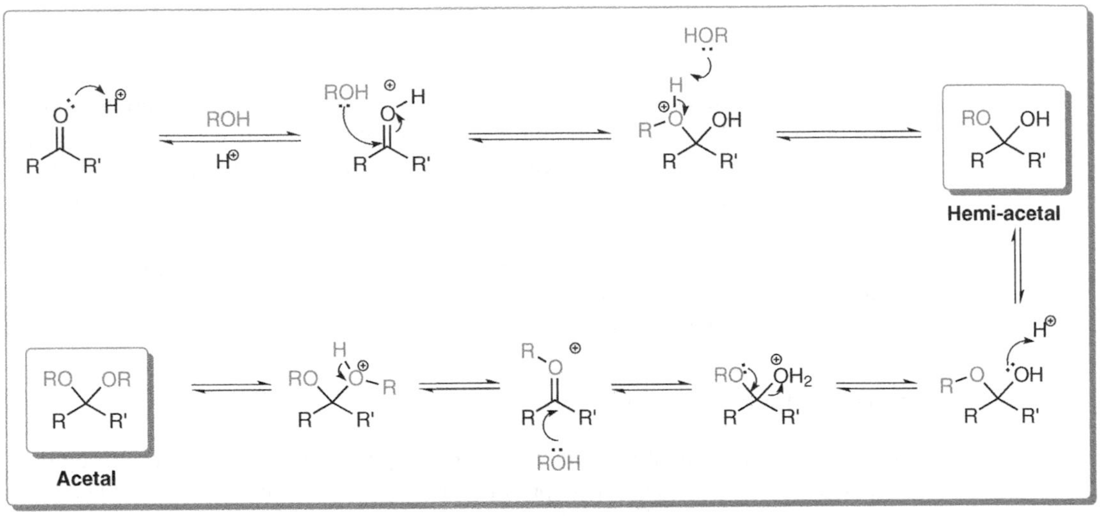

Hemi-acetal

Acetal

Reduction by metal hydrides

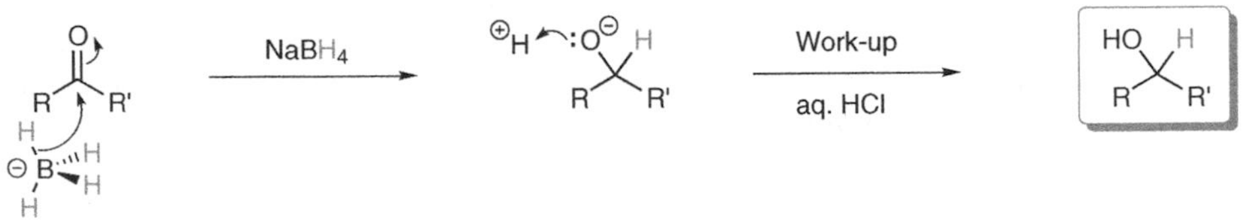

NaBH$_4$

Work-up

aq. HCl

Reduction by hydride transfer

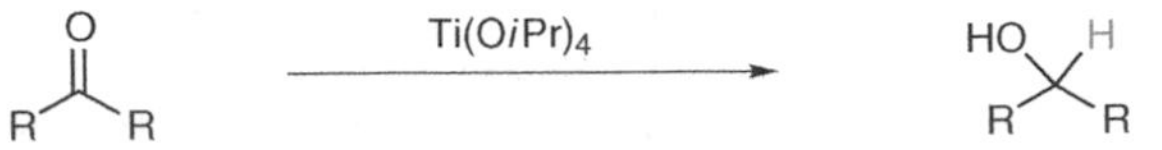

Ti(OiPr)$_4$

via:

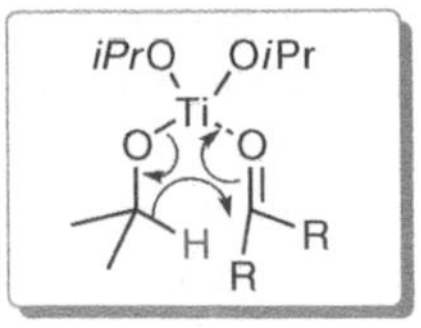

5.41 Nucleophilic addition of heteroatoms to aldehydes and ketones

Hydration

Reaction with water is a reversible process which may be either *general acid* or *general base catalyzed* (see Section 4.5) to form a tetrahedral **hydrate**. Generally, the position of equilibrium favours the carbonyl compound. Increasing the electrophilicity of the carbonyl carbon causes equilibrium to be shifted towards the hydrate. Ketones are reluctant to hydrate unless they possess α-halo substituents, aldehydes hydrate more readily and methanal is almost totally hydrated in water at room temperature.

$\dfrac{[\text{hydrate}]}{[\text{carbonyl}]\,[\text{H}_2\text{O}]}$	$(\text{CH}_3)_2\text{CO}$ 2×10^{-3}	CH_3CHO 1.4	HCHO 2×10^3

Trichloroethanal forms a crystalline, isolable hydrate (*chloral hydrate*).

Hemiacetal formation

This follows a directly analogous pathway to hydration, but in this case the nucleophile is an alcohol. Aldehydes and ketones follow the same reactivity order as in hydration. Hydroxyaldehydes and hydroxyketones in which the hydroxyl group is positioned at C-4 or C-5 form five- and six-membered **cyclic hemiacetals** sometimes referred to as **lactols**. Glucose and fructose are known as **hexoses** (sugars containing six carbon atoms) and are polyhydroxylated aldehydes which can exist in five-ring (**furanose**) and six-ring (**pyranose**) cyclic hemiacetal forms.

Acetal formation

Under acidic conditions, the initially formed hemiacetal can undergo **specific acid catalyzed** (see Section 4.5) loss of water which is assisted by donation of a lone pair of electrons from the adjacent alkoxy-oxygen. Such an effect is termed **neighbouring group participation** or **anchimeric assistance**. This resonance stabilized cationic intermediate may then undergo attack by a second molecule of alcohol and subsequent deprotonation to give the **acetal**, regenerating the proton catalyst. Every step in this sequence is reversible and the position of equilibrium may be controlled by correct choice of reaction conditions.

- Reaction of carbonyl compounds with alcohols under non-aqueous acidic conditions, removing water formed during reaction results in *acetal formation*.
- Reaction of acetals with aqueous acid results in *acetal hydrolysis*.

Due to their stability under basic conditions, acetals are sometimes used as **protecting groups** for aldehydes and ketones which are susceptible to base-catalyzed aldol reactions (see Section 5.51). The carbonyl group can be regenerated by aqueous acid at a later stage.

Thiols undergo an analogous reaction to produce **thioacetals** but more vigorous conditions are required for thioacetal formation and decomposition. Thioacetals are stable under acidic and basic conditions.

Reduction by metal hydrides

Metal hydrides such as LiAlH_4 and NaBH_4 cause reduction of ketones and aldehydes to their corresponding alcohols after hydrolysis of the initial 1,2-adducts.

Reduction by hydride transfer

Metal alkoxides possessing an α-hydrogen may act as sources of hydride and will reduce aldehydes and ketones to alcohols in an equilibrium process. When conditions are set up to result in carbonyl reduction using excess alkoxide reagent, commonly $\text{Ti}(i\text{OPr})_4$, the reaction is known as the **Meerwein–Pondorf–Verley reduction**. The reverse process in which oxidation of the alkoxide is desired occurs if a ketone, such as acetone, is used as the solvent and is referred to as the **Oppenauer oxidation**. Do not forget that these different names apply to the same reaction process which is mechanistically similar to carbonyl reduction by branched Grignard reagents (see Section 5.39).

Aldehydes lacking α-protons may undergo reversible nucleophilic addition of hydroxide and the resulting quaternary alkoxide species may deliver hydride to a second molecule of aldehyde. This reaction, known as the **Cannizzaro reaction**, results in reduction of one molecule and oxidation of the other, a process known as **disproportionation**.

Imine formation

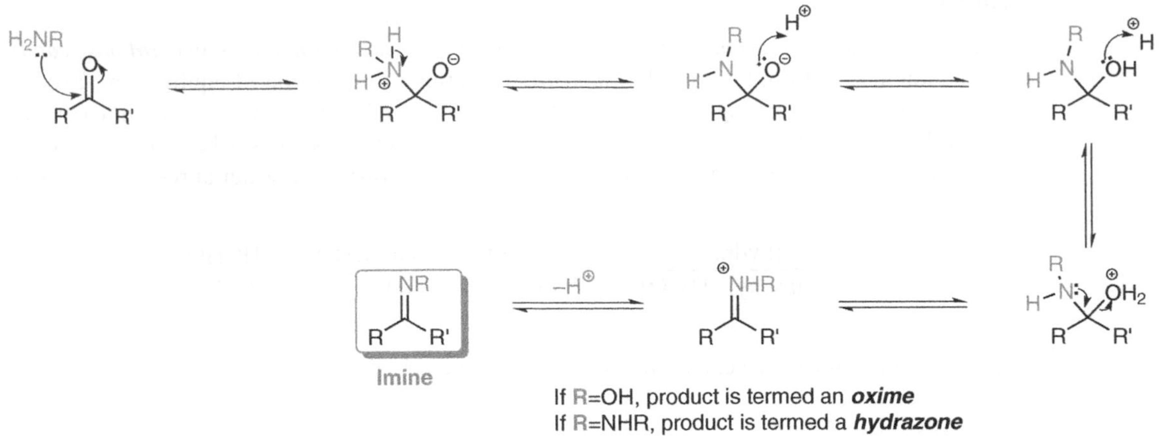

Imine

If R=OH, product is termed an **oxime**
If R=NHR, product is termed a **hydrazone**

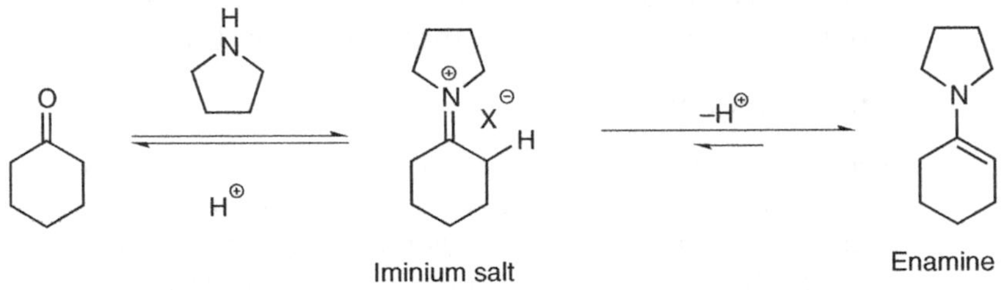

Iminium salt

Enamine

Enol ether formation

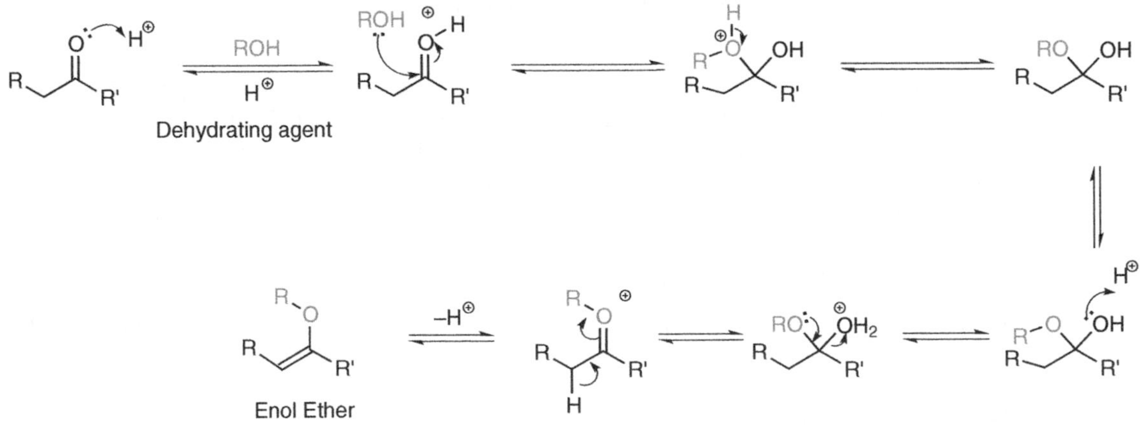

Dehydrating agent

Enol Ether

Tetrahydropyranyl ether (THP ether) formation

R−OH +   →   via:

THP ethers are mixed acetals and are therefore hydrolyzed readily by aqueous acid.

5.42 Condensation reactions of aldehydes and ketones

If the nucleophile is a primary amine (RNH_2) or related structure such as hydroxylamine (NH_2OH) or a hydrazine ($RNHNH_2$), the tetrahedral addition product can undergo elimination of water resulting in *condensation*. The overall sequence is one of *addition–elimination*.

The initial condensation products derived from primary amines are termed *imines* but these may undergo *tautomerism* (see Section 2.18) to the corresponding *enamines* (see Section 2.18). Secondary amines initially form positively charged *iminium salts* which undergo deprotonation to furnish *enamines*.

With unsymmetrical ketones, two regioisomeric enamines are possible but the one possessing the less substituted double bond is often preferred as this suffers less steric congestion.

The condensation products derived from hydroxylamine are termed *oximes* and those from hydrazines are termed *hydrazones*. Highly crystalline oximes and hydrazones such as 2,4-(dinitrophenyl) hydrazones have been used for characterization of carbonyl compounds by virtue of their melting points.

Oximes and hydrazones derived from aldehydes or unsymmetrical ketones display diastereoisomerism but the *E*- and *Z*-isomers may undergo interconversion due to their ability to tautomerize (see Section 2.18). This has significant consequences upon overall stereocontrol of the *Beckmann rearrangement* (see Section 5.59).

5.43 Enol ether formation and reactions

Formation

The acid catalyzed condensation of a carbonyl compound with an alcohol to give an enol ether can be considered to be the oxygen analogue of enamine formation. The reaction pathway is similar to acetal formation (see Section 5.41) up to formation of the intermediate *oxonium ion* but, under the more forcing conditions, proton loss occurs instead of attack by a second molecule of alcohol and an enol ether results.

Reagents such as P_2O_5 and Me_3SiI are used to convert acetals to enol ethers, driving the conversion by reacting with the alcohol formed.

Remember that enol ethers may also be obtained by *O*-alkylation of enolates (see Section 5.47).

Reactions

The double bond of enol ethers is very electron rich and the β-position is strongly nucleophilic. This can be appreciated by a consideration of the extreme resonance canonical forms.

Hydrolysis to the carbonyl compound occurs with aqueous acid and follows the reverse pathway for enol ether formation, being initiated by a rate-determining protonation at the β-position of the double bond. Although the oxygen atom is the most nucleophilic site in the molecule, and reversible protonation occurs most rapidly at this site, this is not a productive pathway.

Tetrahydropyranyl ether formation

Reacting the six-membered cyclic enol ether *dihydropyran* with an alcohol in the presence of an acid catalyst results in the formation of a tetrahydropyranyl ether which is a mixed acetal and can be hydrolyzed by aqueous acid. The ability to mask a hydroxyl group, yet be removed under mild acid conditions, makes dihydropyran a useful reagent for protection of alcohols (see Section 2.8).

Claisen rearrangement of allyl enol ethers

Transetherification of a preformed enol ether with an allylic alcohol under acid catalysis, with heating, results in the formation of an intermediate allyl enol ether. This may undergo a [3,3]-sigmatropic rearrangement (see Section 5.55) to form a carbonyl compound and the overall conversion provides an alternative to enolate alkylation. The driving force for this rearrangement is formation of the thermodynamically more stable carbonyl compound (see Section 5.41).

BAc2 mechanism

$pK_a \approx 4$ $pK_a \approx 16$

AAc2 mechanism

This key step controls the course of the reaction.
Excess water – hydrolysis. Remove water – esterification.

Organometallic reagents with esters

The ketone is more electrophilic and therefore more reactive than the starting ester.

Organolithium reagents

The relative covalency of the O–Li bond and the strong nucleophilicity of Bu⁻ means that attack on the carboxylate is feasible.

Lithium tetrahydroaluminate reduction

74

5.44 Substitution at the carbonyl group in RCOX

With carbonyl substrates RCOX, where X is a *nucleofuge* (see Section 5.1) which can leave as either X⁻ or XH, the initial adduct may collapse to reform the carbonyl group, retaining the nucleophile. This overall *substitution* is a result of sequential *addition-elimination* (see Section 5.19 for another example).

5.45 Ester hydrolysis (saponification)

Seven modes of ester hydrolysis have been described, but the most commonly encountered pathways proceed by either base- or acid-catalyzed addition-elimination mechanisms.

Base-catalyzed bimolecular hydrolysis (BAc2 mechanism)

Rate determining nucleophilic attack of hydroxide onto the carbonyl furnishes a tetrahedral intermediate which collapses, expelling the alkoxide group and generating a carboxylic acid which is converted to its salt under the basic conditions. The acronym **BAc2** refers to the fact that the reaction proceeds via **base**-catalyzed cleavage of the **acyl** carbon–oxygen bond with both hydroxide and ester involved in the rate-determining step.

$$\text{Rate} = k \, [RCO_2R'] \, [HO^-]$$

Acid-catalyzed bimolecular hydrolysis (AAc2 mechanism)

Under acidic conditions, rapid protonation of the carbonyl oxygen precedes rate-limiting nucleophilic attack of water on the carbonyl group. Proton transfer and collapse of the tetrahedral intermediate yields the alcohol and the carboxylic acid. The acronym **AAc2** indicates that the reaction proceeds via acid-catalyzed cleavage of the acyl carbon–oxygen bond with both water and ester involved in the rate-determining step.

$$\text{Rate} = k \, [RCO_2R'] \, [H_2O]$$

This process is fully reversible and the equilibrium position can be controlled by choice of reaction conditions.

- Reaction of a carboxylic acid and an alcohol in the presence of an acid catalyst, with removal of water, leads to *condensation* resulting in *esterification*.
- Reaction of an ester with aqueous acid results in *hydrolysis*, sometimes referred to as *saponification*.

5.46 Special cases

Organometallic reagents with esters

When esters react with Grignard reagents, the initial substitution product is a carbonyl derivative which is more reactive than the starting ester (see Section 5.37). A second nucleophilic addition produces an alcohol in which two of the substituents are derived from the organometallic reagent.

Organolithium reagents with carboxylic acids

Carboxylic acids generally do not undergo substitution under basic conditions as they are simply deprotonated to form unreactive carboxylate salts. However, the high nucleophilicity of organolithium reagents, coupled with the significantly covalent nature of the O–Li bond, results in subsequent 1,2-addition to give a stable tetrahedral adduct. On hydrolysis, this adduct decomposes to a ketone. Note that the ketone is only generated on work-up, subsequent to destruction of any remaining organolithium reagent. Grignard reagents are insufficiently nucleophilic to undergo this reaction.

Lithium tetrahydroaluminate (LiAlH₄) reduction of esters and amides

Initial reduction of an ester proceeds by 1,2-addition to the carbonyl group and subsequent collapse of the tetrahedral intermediate expelling alkoxide. This yields an aldehyde which is more electrophilic than the ester (see Section 5.37) and therefore more reactive towards reduction. A second reduction converts the aldehyde to a primary alcohol.

Amides similarly undergo carbonyl 1,2-addition, but the tetrahedral intermediate collapses with expulsion of the oxygen substituent which is a better nucleofuge than the amine group. The imine or iminium species produced then undergoes further reduction by 1,2-addition, liberating an amine after aqueous work-up.

Enolate structure

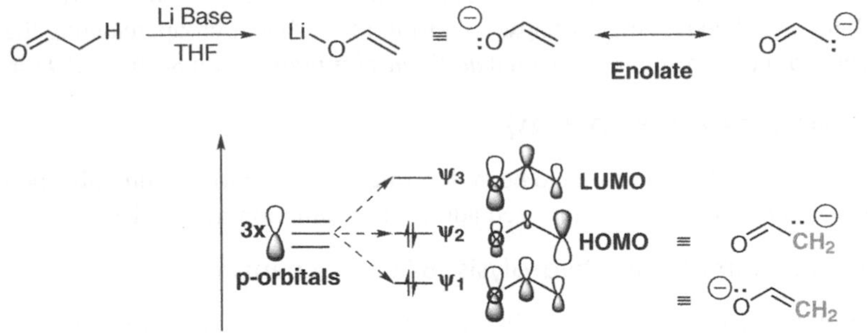

Lithium resides mainly on oxygen

Enolate

LUMO

HOMO

$3x$ **p-orbitals**

Reactivity

LDA =

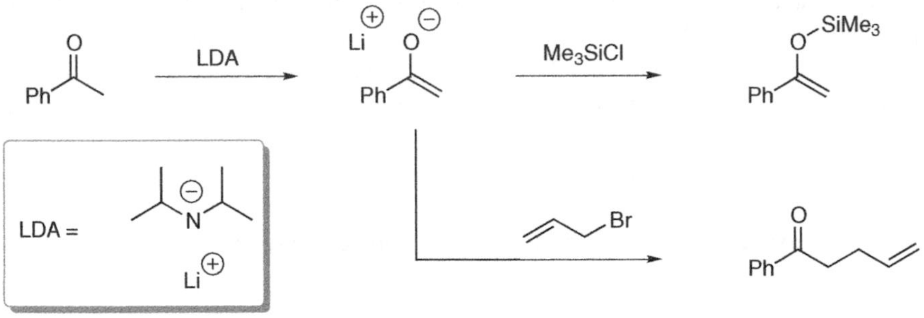

Hard Electrophile

adds to the hard nucleophilic centre of the enolate

Soft Electrophile

adds to the soft nucleophilic centre of the enolate

Kinetic versus thermodynamic enolates

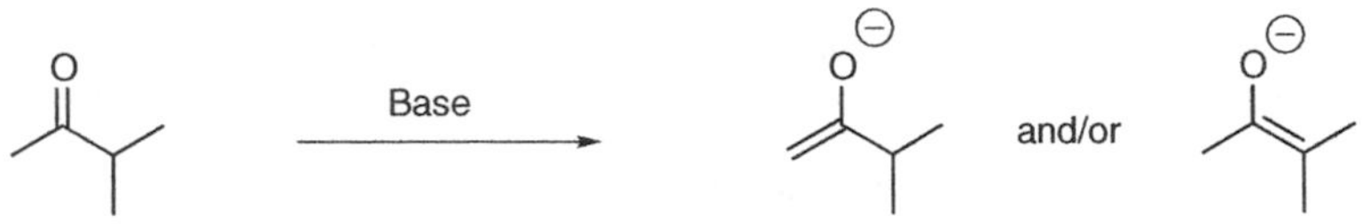

Kinetic and/or *Thermodynamic*

Two possible enolates, how can selectivity be achieved?

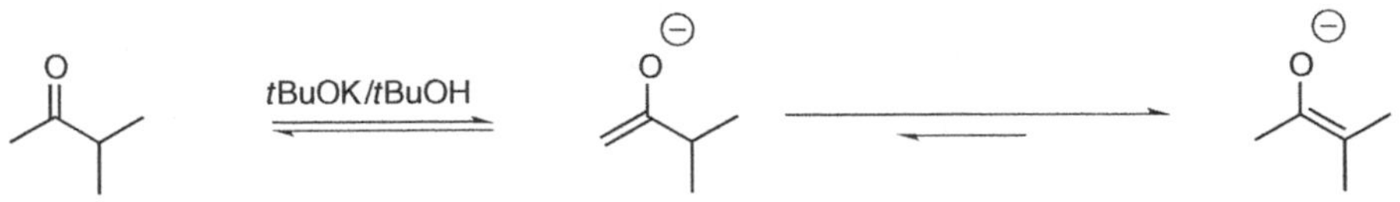

tBuOK/tBuOH

Thermodynamic enolate

LDA

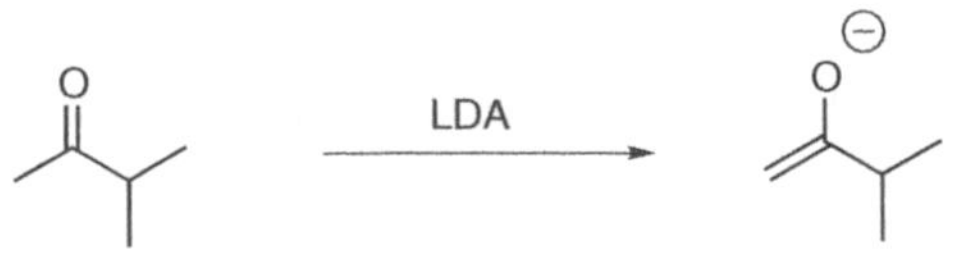

Kinetic enolate

5.47 Enolate chemistry

Enolate structure

Deprotonation at positions α- to a carbonyl group occurs more readily than in simple hydrocarbons because the anionic species produced is stabilized by *resonance* (see Section 2.12). This *enolate* species is an *ambident anion* which can be represented by two extreme resonance forms in which the negative charge is located on carbon or oxygen. The higher electron density resides on oxygen due to its greater electronegativity (see Section 2.11).

> - Always use the extreme resonance form with the negative charge on oxygen when drawing reaction mechanisms involving enolates, but remember that both carbon and oxygen are nucleophilic.

Reaction of enolates with electrophiles: reaction at oxygen versus carbon

According to *HSAB* terminology, the enolate oxygen is the *hard nucleophilic centre* and the carbon is the *soft nucleophilic centre*. Species reacting as hard electrophiles (e.g. Me_3SiCl) tend to react at the oxygen; whereas those reacting as soft electrophiles (e.g. $PhCH_2Br$, $CH_2=CHCH_2Br$) predominantly react at carbon.

Alkylation at the α-position of carbonyl compounds is a useful synthetic transformation but *O*-alkylation always competes and elimination from the electrophile may be problematic.

Phenolates are special cases of enolates in which negative charge resides almost totally on the oxygen due to the aromatic nature of the aryl ring and electrophiles react almost exclusively on the oxygen.

Increasing ionic nature (hardness) of the enolate results in increased reaction at oxygen. The inverse favours reaction at carbon. This can be affected by the *dielectric constant* (ε) of the solvent (polarity) and the electropositive nature of the counterion. Polar solvents and highly electropositive counterions favour ionic enolates; whereas low polarity solvents and less electropositive counterions favour more covalent enolates.

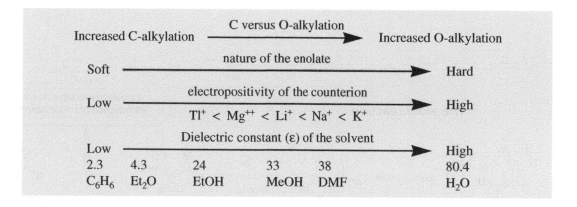

5.48 Kinetic versus thermodynamic enolates

Deprotonation of unsymmetrical ketones can result in formation of regioisomeric enolates. Choice of reaction conditions may permit selective generation of either the kinetic or the thermodynamic enolate (see Section 4.3).

Kinetic enolates arise from removal of one of the most accessible, acidic and abundant α-protons under non-equilibrating conditions (low temperature, short reaction times, strong base).

Thermodynamic enolates have the most substituted double bond and are generated under equilibrating conditions (higher temperature, longer reaction times and milder base).

> With simple ketones:
>
> - Deprotonation of the less substituted α-carbon forms the *kinetic enolate*.
> - Deprotonation of the more substituted α-carbon forms the *thermodynamic enolate*.

Stork enamine reaction

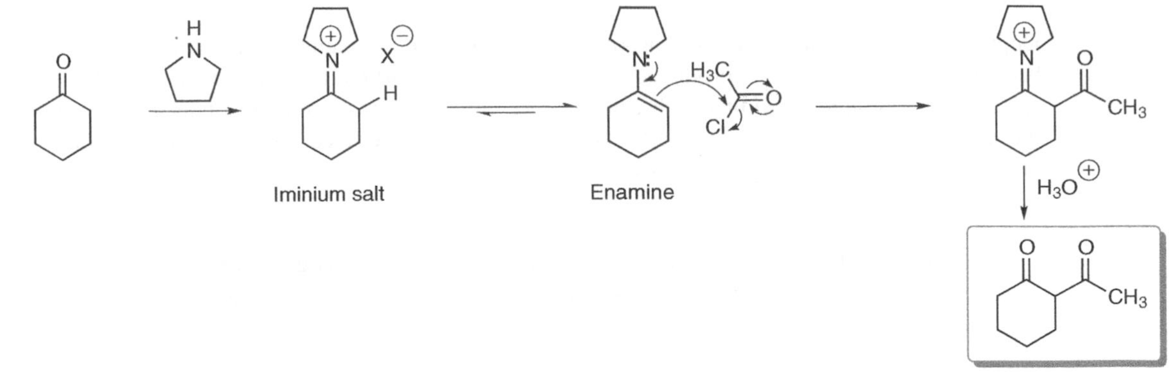
Iminium salt
Enamine
$H_3O^{\oplus}$

Acetoacetic/malonic ester synthesis

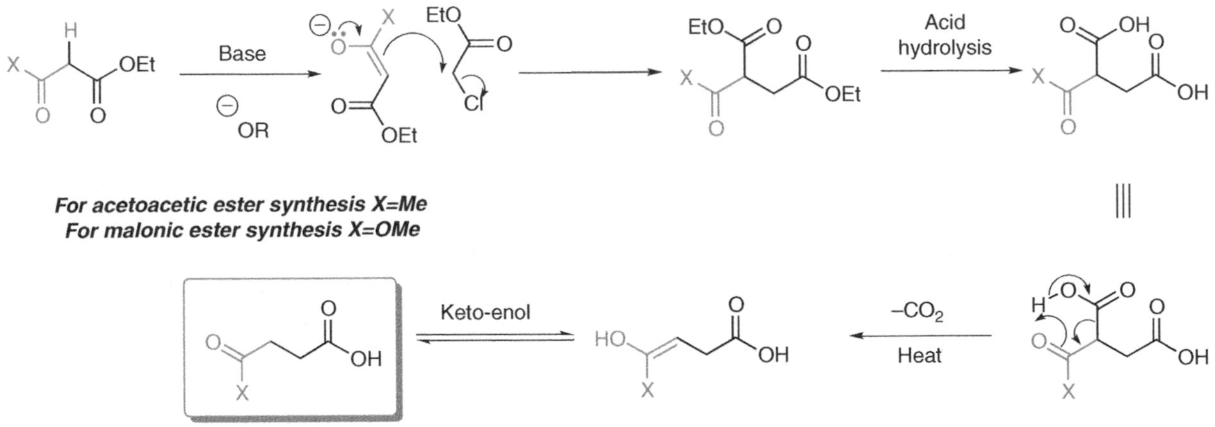
Base
Acid hydrolysis
For acetoacetic ester synthesis X=Me
For malonic ester synthesis X=OMe
Keto-enol
$-CO_2$
Heat

The aldol reaction

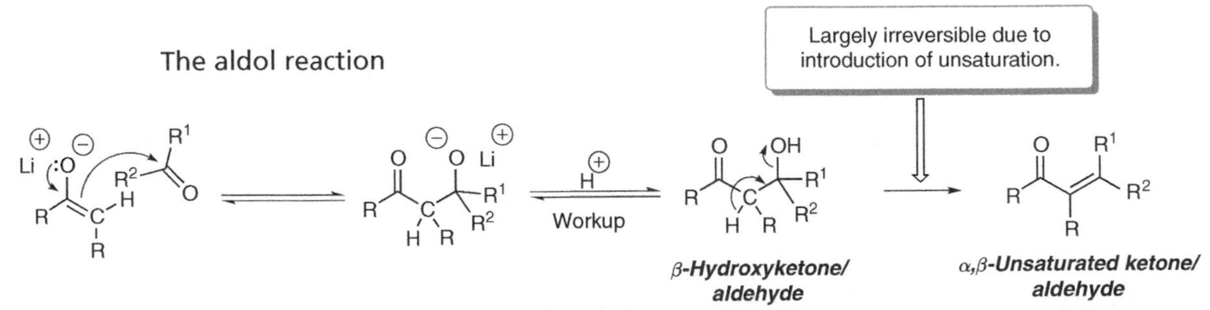
Largely irreversible due to introduction of unsaturation.
Workup
β-Hydroxyketone/ aldehyde
α,β-Unsaturated ketone/ aldehyde

β-Elimination is unimolecular and proceeds via the conjugate base

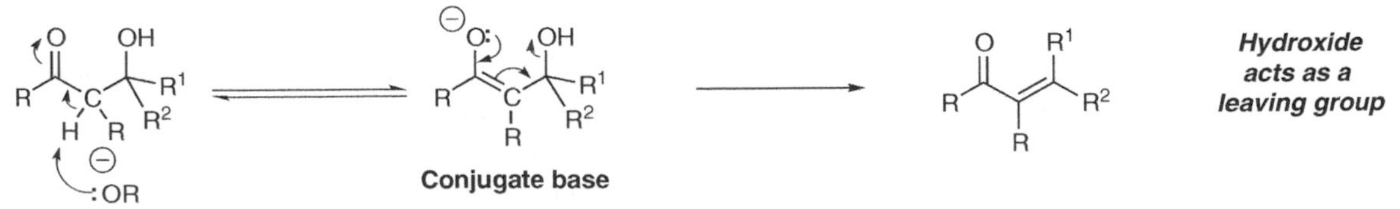
Conjugate base
Hydroxide acts as a leaving group

Acidic conditions result in efficient elimination

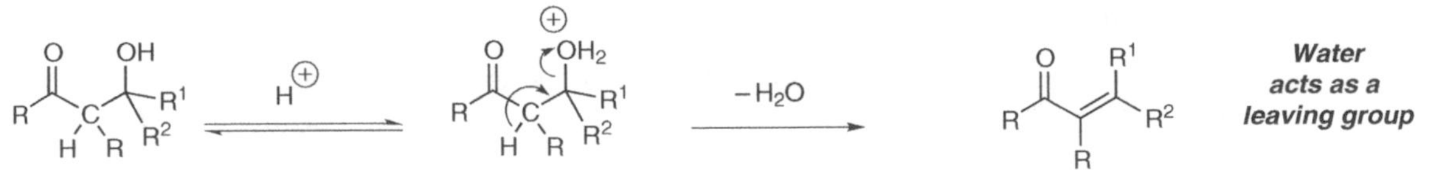
$-H_2O$
Water acts as a leaving group

5.49 The Stork enamine synthesis

By analogy with enol ethers, enamines have an electron-rich double bond and are nucleophilic at the β-carbon (see Sections 5.43). Their reactivity is intermediate between the parent carbonyl and its enolate and they will undergo 1,4-addition to α,β-unsaturated carbonyl compounds (see Section 5.40). In addition, enamines undergo acylation and alkylation at the β-carbon.

Aqueous hydrolysis of the products of these reactions regenerates the carbonyl group, thus providing a milder alternative to enolate generation and reaction.

5.50 Acetoacetic ester and malonic ester syntheses

Ethyl 3-oxobutanoate (ethyl acetoacetate) and diethyl propandioate (diethyl malonate) occupy key positions in synthetic enolate chemistry because of their special reactivity. In both cases, the 1,3-relationship of the carbonyl groups renders the methylene group appreciably more acidic than in simple carbonyl compounds due to co-operative stabilization of the enolate. In addition, their derived carboxylic acids (β-*ketoacids*) readily undergo thermally induced loss of carbon dioxide (*decarboxylation*), the overall sequence providing a milder alternative to enolate generation and reaction.

Acetoacetic ester synthesis

Alkylation of ethyl 3-oxobutanoate followed by acid hydrolysis and decarboxylation results in products which are formally derived by alkylation of the enolate derived from propanone. However, the increased acidity of ethyl 3-oxobutanoate (pK_a 10.7) compared to propanone (pK_a 20.0) results in improved yield by increasing the proportion of *C*-alkylation and decreasing polymerization due to *aldol reactions* (see Section 5.51). This procedure is often referred to as the *acetoacetic ester synthesis*.

Malonic ester synthesis

In the same manner, diethyl propandioate (pK_a 13.3) can be alkylated, hydrolyzed and decarboxylated to give products formally derived from methyl ethanoate (pK_a 24) and this procedure is known as the *malonic ester synthesis*.

5.51 The aldol reaction

Generation of an enolate from an aldehyde or ketone with a base of similar pK_a (for instance sodium ethoxide) results in an equilibrating system in which both enolate and parent carbonyl compound are present. Under such conditions, the nucleophilic enolate reacts at the electrophilic carbon of the carbonyl group of the un-ionized partner. This results in C–C bond formation and subsequent protonation leads to a β-hydroxycarbonyl adduct.

A trivial name for the product arising from the base catalyzed dimerization of ethanal is *aldol* (*ald*ehyde + alcoh*ol*) and the name *aldol reaction* is now applied to all such reactions of aldehydes and ketones. Self condensation of aldehydes and symmetrical ketones gives a single aldol product; whereas unsymmetrical ketones and reactions involving more than one carbonyl compounds give rise to mixtures of products. The latter case is termed the *crossed aldol reaction* and it is sometimes possible to control reaction conditions in order to favour a single product.

Under more forcing conditions the initial aldol product may dehydrate forming the thermodynamically more stable α,β-unsaturated carbonyl compound.

1,4- and 1,5-dicarbonyl compounds may undergo intramolecular aldol reaction, followed by dehydration to produce conjugated cyclopentenones and cyclohexenones respectively. In a specific example, known as the *Robinson annelation*, the enolate of a cyclic 1,3-diketone undergoes initial *Michael addition* (see Section 5.40) to a vinyl ketone and the resultant adduct cyclizes by intramolecular aldol reaction, followed by dehydration to give a fused cyclohexenone product.

The aldol reaction is the archetype of a family of related transformations in which an anion derived from an acidified alkyl group attacks a carbonyl substrate (see next section).

Claisen–Schmidt reaction

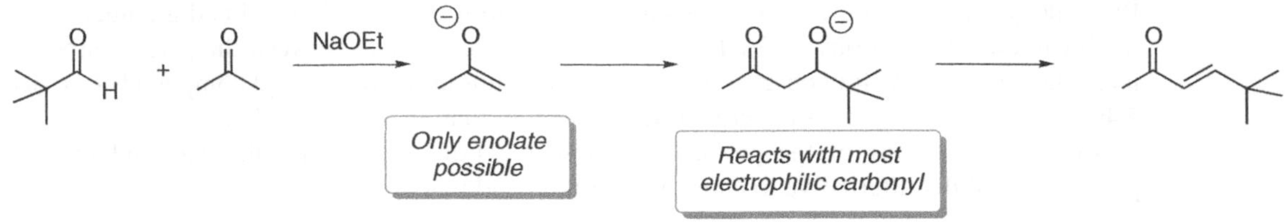

Only enolate possible

Reacts with most electrophilic carbonyl

Mannich reaction

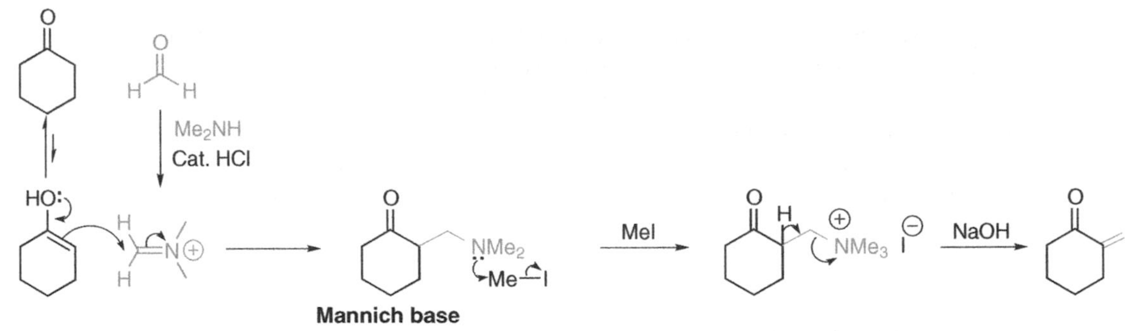

Mannich base

Stobbe condensation

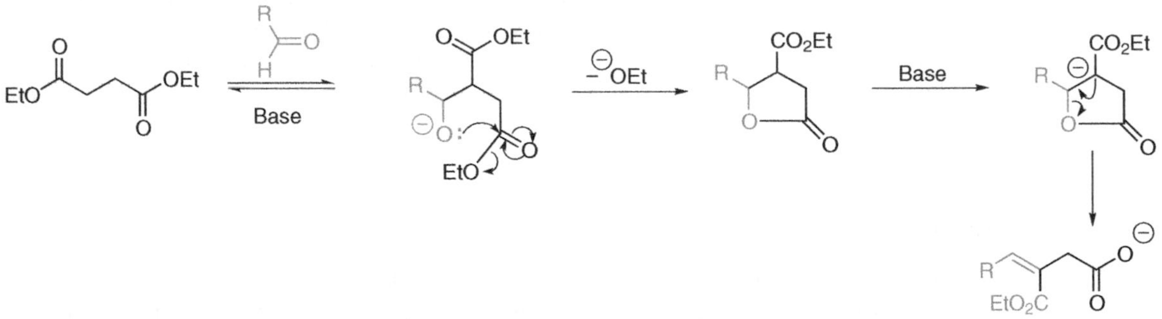

Knoevenagel reaction

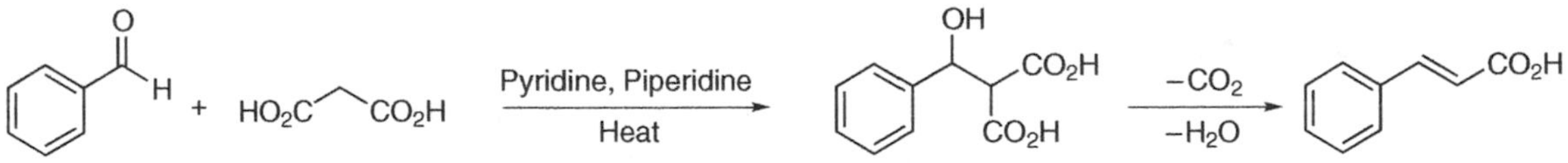

Darzens glycidic ester condensation

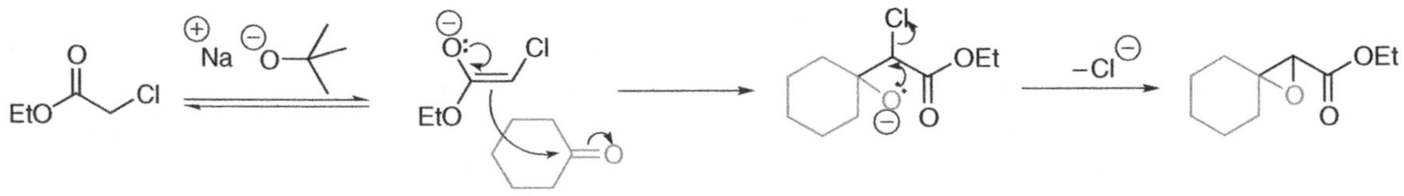

Reformatsky reaction

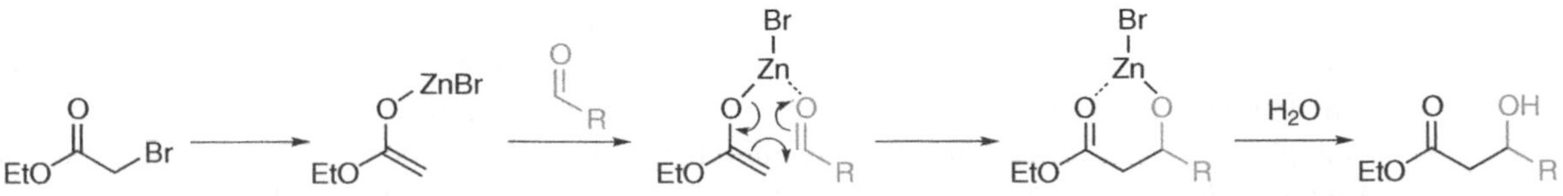

The zinc enolate is insufficiently reactive to attack another ester molecule. It will only react with a more electrophilic carbonyl compound.

5.52   Variants of the aldol reaction

Claisen–Schmidt reaction

This is an example of a crossed aldol reaction between an aldehyde possessing no α-hydrogens and a symmetrical ketone. The success of this reaction results from the greater reactivity of the aldehyde carbonyl towards nucleophilic attack (see Section 5.36) and the fact that only the ketone can form an enolate.

Mannich reaction

The standard crossed aldol condensation is not suitable when one of the components is methanal as the resultant terminal vinyl ketones undergo anionic polymerization under the basic reaction conditions. The **Mannich reaction** is a two-step process avoiding the use of strongly basic conditions. A ketone and methanal are mixed in the presence of a buffered secondary amine. The more electrophilic methanal reacts preferentially with the amine to form a highly electrophilic **iminium** species which is attacked by the enol tautomer of the ketone to form the α-(dialkylamino)methyl addition product.

Stobbe reaction

Butan-1,4-dioate esters (succinate esters) on treatment with alkoxide bases form a monoenolate which reacts with aldehydes to give an alkoxy intermediate which undergoes γ-**lactonization** by intramolecular attack on the ester group situated five atoms away. However, the lactone is not the final product as it undergoes deprotonation followed by irreversible eliminative ring opening to generate an α,β-unsaturated ester with an ethanoate side chain. In the overall condensation, one of the two ester groups of the starting material is specifically hydrolyzed.

Knoevenagel reaction

This is another example of a crossed aldol reaction in which an aldehyde lacking α-hydrogens reacts with an enolate generally derived from compounds possessing a methylene or methine group attached to two mesomerically electron-withdrawing substituents (ZCHRZ′, where Z and Z′ may be CO_2H, CO_2R, COR, CHO, CN, NO_2). With the exception of Z=CO_2H, the activated methylene substrates undergo addition followed by dehydration to give the condensation products. If one of the activating groups is a carboxylic acid however, the initial aldol adduct usually undergoes spontaneous decarboxylation and dehydration to produce a disubstituted alkene. A range of bases may be used to catalyze the reaction although secondary amines such as piperidine are commonly used. Frequently a mixture of piperidine and pyridine is used and this protocol is referred to as the **Doebner modification** of the Knoevenagel reaction.

Darzens glycidic ester condensation

The enolate derived from an α-haloester will undergo a normal aldol reaction with an aldehyde or ketone, but the intermediate alkoxide displaces halide by an intramolecular S_N2 reaction to give an α,β-epoxy ester (**glycidic ester**).

Reformatsky reaction

Reaction of α-haloesters with activated zinc results in formation of an organozinc reagent which behaves like the analogous enolate but with a diminished nucleophilicity. This reagent reacts selectively with aldehydes and ketones but not with the starting ester. Aqueous work-up usually produces the β-hydroxyester aldol product without subsequent dehydration occurring.

Claisen condensation

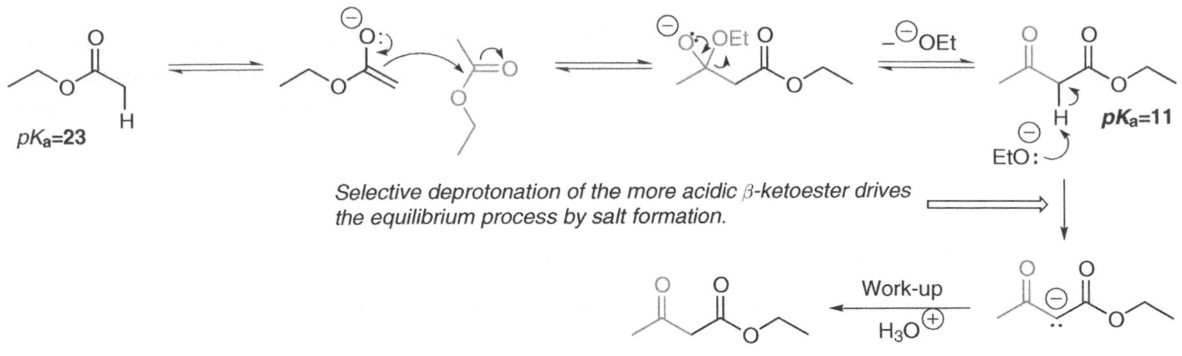

pKa=23
OEt
pKa=11
EtO:
Selective deprotonation of the more acidic β-ketoester drives the equilibrium process by salt formation.
Work-up
H3O+

Wittig reaction

B:
Base
Bond rotation
betaine
The driving force for the process is the strength of the P=O bond (540 kJMol−1)
oxaphosphetane
PPh3
EtO2C
Stabilized ylide
Me
Non-stabilized ylide

Horner–Emmons (Wadsworth–Emmons) reaction

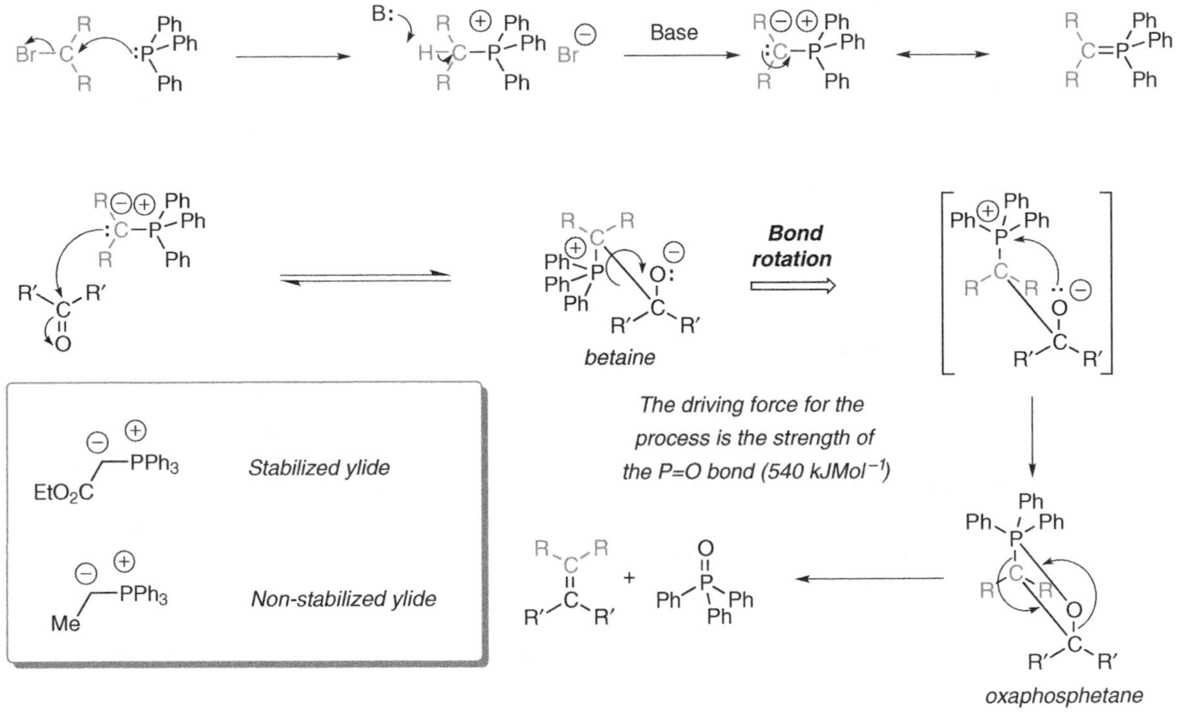

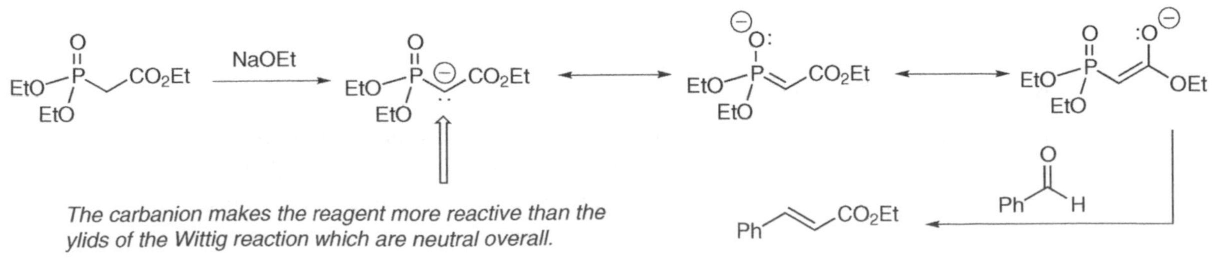

NaOEt
CO2Et
EtO
OEt
The carbanion makes the reagent more reactive than the ylids of the Wittig reaction which are neutral overall.

5.53 Base catalyzed self condensations of esters

Claisen condensation

α-Deprotonation of an ester with alkoxide generates a low equilibrium concentration of enolate which will react with the excess of ester present to give a tetrahedral intermediate. This may collapse, with expulsion of alkoxide to generate a *β-ketoester* and so, in principle, only a catalytic amount of alkoxide is necessary for reaction. However, the equilibrium for this process lies well towards starting materials and, in order to force reaction to completion, at least one equivalent of base is necessary. The β-ketoester possesses a more acidic methylene group than the starting ester as the anion is stabilized mesomerically by two carbonyl groups (pK_a ester ~25, pK_a β-ketoester ~11). It is fully deprotonated by the alkoxide (pK_a ~16), precipitating from solution as its salt and thus displacing equilibrium towards product. β-Ketoesters are useful starting materials for the *acetoacetic ester synthesis* (see Section 5.50). The intramolecular variant is known as the *Dieckmann reaction* and is successful for five-, six- and seven-membered rings, intermolecular reaction dominates with other ring sizes.

5.54 Reactions of phosphorus stabilized carbanions

Wittig reaction

Triarylphosphines are nucleophilic species and will react with alkyl halides to form *phosphonium salts*. This reactivity is similar to that of tertiary amines which form quaternary ammonium salts. However, unlike nitrogen, phosphorus may accept a pair of electrons into a low energy d-orbital, thus increasing the outer shell to 10 electrons. The consequence of this is that deprotonation of a phosphonium salt α- to phosphorus is facilitated, generating a neutral resonance stabilized *phosphorus ylide* or *phosphorane* which is sufficiently nucleophilic to react with the carbonyl groups of aldehydes and some ketones.

In the generally accepted mechanism for the Wittig reaction, the initial intermediate is a dipolar *betaine* which then closes to an *oxaphosphetane*. Finally, extrusion of a phosphine oxide generates an alkene. The driving force for this exothermic reaction is the strength of the P=O bond which is approximately 540 kJ mol^{-1}.

Ylides fall into two distinct categories. *Stabilized ylides* possess a mesomerically electron withdrawing group (e.g. ester, nitrile, carbonyl) on the α- carbon; whereas *non-stabilized* ylides possess no such functionality. These two types of ylide frequently result in stereocomplementarity in the geometry of the product alkenes; stabilized ylides favouring formation of the *E*-alkene and non-stabilized ylides favouring the *Z*-alkene. This may be rationalized on the basis of reversibility or otherwise in the betaine formation.

With stabilized ylides the betaine which leads to the *anti*-oxaphosphetane reacts faster than its diastereoisomer which may instead revert to starting materials. Subsequent *syn*- elimination of the phosphine oxide from the *anti*-oxaphosphetane leads to the *E*-alkene. Thus it is the relative thermodynamic stability of the oxaphosphetanes derived from stabilized ylides which determines the product distribution (see Section 4.3).

Non-stabilized ylides react irreversibly with the carbonyl group and so it is the most readily formed oxaphosphetane which results in the major alkene isomer. One hypothesis invokes preferred Bürgi–Dunitz approach (see Section 5.38) by the non-stabilized ylide such that the phosphonium substituent is furthest away from the bulkier of the two carbonyl substituents. Cyclization of the predominant betaine leads to the *syn*-oxaphosphatane and then to the *Z*-alkene. The ultimate product composition is therefore under kinetic control (see Section 4.3).

Horner–Emmons (Wadsworth–Emmons) reaction

This is a related reaction wherein a carbanion is stabilized by a phosphonate ester and shows two advantages over the Wittig reaction. The enhanced nucleophilicity of the carbanionic species means that ketones as well as aldehydes can be used as substrates. In addition, the water solubility of the dialkyl phosphate ester by-product results in a much easier isolation procedure for the alkene. The reaction mechanism is similar to that of the Wittig reaction.

Pericyclic reactions – interaction diagrams

Transition state	Transition state	Transition state
[4+2] cycloaddition	Hexatriene cyclization	Hexatriene cyclization
Hückel array	Disrotatory closure	Conrotatory closure
Aromatic	Hückel array	Möbius array

Frontier molecular orbital analysis

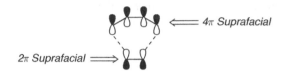

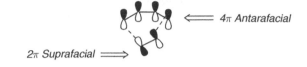

Cycloadditions

2 +1 Cycloaddition

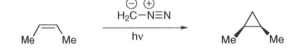

Stereospecific carbene addition.
Geometry of double bond preseved
in cyclopropane.

Frontier orbital description

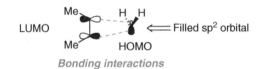

2 +2 Cycloaddition

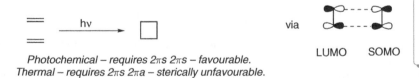

Photochemical – requires $2\pi s\ 2\pi s$ – favourable.
Thermal – requires $2\pi s\ 2\pi a$ – sterically unfavourable.

An Electron is excited from the HOMO to the next lowest-energy orbital, this is known as the SOMO. Reaction proceeds via interaction of the SOMO with the LUMO of another alkene

Ketenes

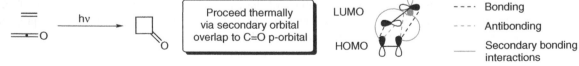

Proceed thermally via secondary orbital overlap to C=O p-orbital

---- Bonding
---- Antibonding
___ Secondary bonding interactions

Paterno–Büchi reaction

Oxetane formation

via diradical intermediate

5.55 Pericyclic reactions

Pericyclic reactions are defined as *concerted processes* which pass through a cyclic transition state in which π- and σ-bonds are made and broken simultaneously (although many examples are known which exhibit some degree of asynchronous character). Pericyclic reactions may be grouped into four categories: *cycloadditions, electrocyclizations, sigmatropic rearrangements* and *cheletropic reactions*. Reactions are defined as *allowed* or *forbidden* depending upon electronic constraints and display a high level of stereocontrol under defined reaction conditions. The stereochemical outcome may be predicted by consideration of symmetry elements of the molecular orbitals of reactants and products.

One approach towards rationalizing reactivity applies *selection rules* based on the premise that ground state (thermal) pericyclic reactions involving $4n+2$ electrons occur through aromatic-like transition states; whereas excited state (photochemical) reactions proceed via anti-aromatic-like transition states. This situation is reversed for systems involving $4n$ electrons.

Pericyclic reactions may follow two distinct stereochemical pathways. If bonds are made and broken on a single face of a reactant, the interaction is described as *suprafacial* (s) with respect to that component. If both faces of a reactant are involved in bond making and breaking the interaction is described as *antarafacial* (a). If all components interact in a suprafacial manner the electrons in the transition state make up a *Hückel array*, but if one component is antarafacial, the interaction is termed a *Möbius array*. In this Hückel–Möbius analysis, the selection rules simply state:

- *Concerted pericyclic reactions involving $4n+2$ electrons occur through a Hückel transition state under thermal conditions and a Möbius transition state under photochemical conditions.*
- *Concerted pericyclic reactions involving $4n$ electrons occur through a Möbius transition state under thermal conditions and a Hückel transition state under photochemical conditions.*

Whatever the prediction on electronic grounds, the reaction may still not occur due to steric constraints. Antarafacial processes are often so disfavoured that many pericyclic reactions only take place under conditions that allow suprafacial interactions. Examples of some synthetically important pericyclic reactions are listed here.

Cycloadditions

Two reacting components are involved in the simultaneous formation of two σ-bonds across the termini of each system, generally resulting in ring formation.

2+1 Cycloaddition

Singlet carbene generated under photochemical conditions undergoes concerted suprafacial cycloaddition with alkenes to generate cyclopropanes in which the stereochemistry of the alkene is reflected in the relative stereochemistry of the cyclopropane product.

2+2 Cycloaddition

Cycloaddition between two alkenes is a 4π-electron process and would require one of the components to orient in an antarafacial manner for the reaction to occur thermally ($2\pi s + 2\pi a$). This incurs high steric barriers and so the reaction only occurs under photochemical (excited state) conditions when suprafacial interaction of both components is the allowed pathway ($2\pi s + 2\pi s$). This reaction is by far the most general means of constructing cyclobutanes.

Ketenes can undergo thermal $2\pi s + 2\pi a$ cycloadditions as the linear ketene can permit an antarafacial orientation to its partner. Thermal dimerization of ketene leads to the formation of a β-lactone.

The variant in which one of the 2π components is a carbonyl group is called the *Paterno–Büchi* reaction. This may occur either by a photochemically induced concerted cycloaddition or a thermal stepwise pathway to form an *oxetane*.

3 +2 Dipolar cycloaddition

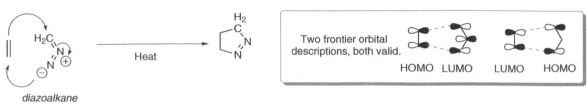

diazoalkane

Other 1,3 dipoles

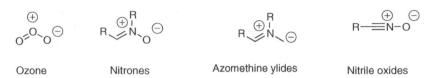

Ozone   Nitrones   Azomethine ylides   Nitrile oxides

Diels–Alder reaction

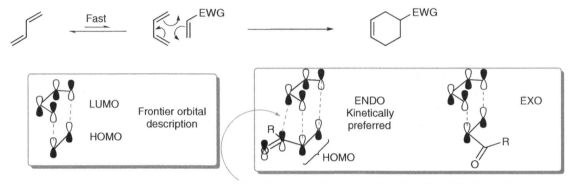

Positive non-bonding interaction. Stabilizes TS, absent in product

Ene reaction

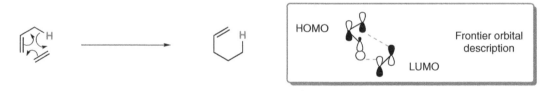

Electrocyclizations

Butadiene–cyclobutene rearrangement

Butadiene–cyclobutene rearrangement

Hexatriene–cyclohexadiene rearrangement

3 +2 Dipolar cycloaddition

The 2π component (**dipolarophile**) is usually an alkene but other π-systems such as carbonyl groups or imines can also be involved. The 4π partner (**dipole**) necessarily exists as an overall neutral, but charge separated, **ylide** which can be represented by several resonance canonical forms. Although all resonance canonicals are equivalent (see Section 2.12), for mechanistic considerations it is preferable to consider the canonical in which the central atom is positively charged and the negatively charged atom has fewer bonds to the central atom than the neutral centre. Because 6π electrons are involved, the cycloadditions occur thermally and entail suprafacial interaction of both components.

Dipolar species may be isolable (**ozone, nitrones, diazoalkanes**) or may be generated as reactive intermediates (**azomethine ylides, nitrile oxides**).

Diels–Alder reaction

This reaction, discovered in 1928 by Otto Diels and Kurt Alder, for which they were awarded the Nobel Prize in 1950, involves thermal reaction between an 1,3-diene and an alkene (**dienophile**) to form a cyclohexene. Generally, electron donating substituents on the diene and electron withdrawing substituents on the dienophile favour reaction. The regiochemical outcome is influenced by the polarization of the reactants. Lewis acids (see Section 1.4) may catalyze reaction and amplify regiocontrol by complexing with the dienophile, further withdrawing electrons and enhancing polarization.

As 6π electrons are involved, this reaction occurs under thermal conditions and both components interact suprafacially. This requires the dienophile to approach above or below the plane of the diene, with the diene in an s-**cis** *conformation*. If the electron withdrawing substituent on the dienophile is disposed away from the diene, the stereochemistry of approach is defined as **exo**, if towards it, **endo**. Although exo- products would be expected for steric reasons, endo- approach is kinetically preferred due to favourable non-bonding interactions in the transition state and endo-cycloadducts usually predominate (**kinetic endo effect**).

This is one of the most powerful reactions available to the synthetic chemist, resulting in regiocontrolled formation of a functionalized six-membered ring with stereochemical control over the relative configurations of the sp^3 centres.

Ene reaction

The reaction between an alkene (**enophile**) and an allylic species involves formation of one carbon–carbon σ-bond and the transfer of a hydrogen atom from the allylic fragment to the enophilic partner. This results in the formation of an acyclic product with a pentenyl core. Factors which accelerate or inhibit the Diels–Alder reaction usually act in a similar manner in the ene reaction.

Electrocyclizations

Suprafacial reaction involves ring closure by the interacting orbitals involved in ring formation or cleavage rotating in opposite senses (**disrotatory**); whereas antarafacial reaction requires both orbitals to rotate in the same sense (**conrotatory**). The disrotatory process corresponds to an aromatic Hückel-type of interaction; whereas the conrotatory process corresponds to an anti-aromatic Möbius interaction. *Note*: There are two possible disrotatory or conrotatory modes possible for every substrate and which is followed may be directed by steric requirements of the reactant.

Butadiene–cyclobutene rearrangement

Thermal treatment of cyclobutenes can result in conrotatory reverse electrocyclization to butadiene derivatives. Conversely, irradiation of butadienes at a wavelength absorbed only by the diene can result in disrotatory electrocyclization to form the cyclobutene.

Hexatriene–cyclohexadiene rearrangement

Thermal treatment of cyclohexadienes can result in disrotatory reverse electrocyclization; whereas photochemical conditions result in a conrotatory process.

[1,3]-sigmatropic rearrangement

H 1s Thermal reaction
HOMO
Antarafacial migration
Geometrically impossible

H 1s Photochemical reaction
SOMO
Suprafacial migration
Geometrically possible

[1,5]- and [1,7]-sigmatropic rearrangements

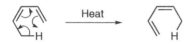

Molecular orbital depiction
for thermal 1,5-shift.
Suprafacial
LUMO diene
HOMO C–H σ-bond

LUMO HOMO

Thermal reaction results in antarafacial H transfer

[3,3]-sigmatropic rearrangement

Cope

Oxy-Cope

H

Claisen

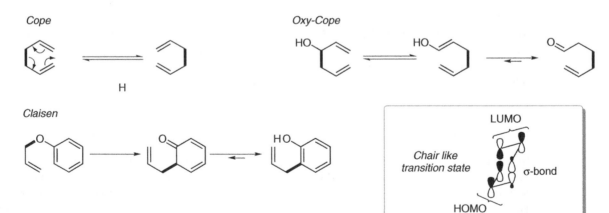

*Chair like
transition state*

LUMO

σ-bond

HOMO

Cheletropic reactions

Sigmatropic rearrangements

As the term *sigmatropic* implies, a group migrates around a conjugated polyene by making and breaking of σ-bonds at the termini. Rearrangements are classified according to the number of atoms constituting each of the formal fragments (e.g. [1,3]-, [3,3]-, [1,5]-). For most systems, steric constraints generally mean that the migrating group can only migrate suprafacially over the polyene framework. If the migrating group is hydrogen, the nature of the hydrogen s-orbital means that it too must contribute a suprafacial component and therefore only a single reaction condition (thermal or photochemical) permitting suprafacial migration exists for each [1,n]-rearrangement. However, carbon may migrate either suprafacially, retaining the configuration of the migrating centre, or antarafacially, inverting the configuration of the migrating centre.

[1,3]-Sigmatropic rearrangements

Involving 4π electrons, the thermal reaction requires an antarafacial component to the transition state. This is sterically impossible for hydrogen migrations and so these reactions only occur photochemically. *Note*: It is for this reason that keto-enol tautomerism proceeds by a stepwise mechanism involving protonation-deprotonation and not via a concerted rearrangement.

Photochemical and thermal carbon migrations may occur, the latter involving an antarafacial component in the migrating carbon which results in inversion of stereochemistry at this centre.

[1,5]-Sigmatropic rearrangements

Suprafacial [1,5]-hydrogen migration occurs readily and is the reason for the ready interconversion of 3-substituted cyclopentadienes and their 1-substituted isomers.

[1,7]-Sigmatropic rearrangements

Migrations over extended π-frameworks are less common but an example of medical significance occurs in the sunlight promoted conversion of procalciferol (available from ergosterol in the diet) to calciferol (vitamin D_2) by a suprafacial 1,7-hydrogen migration. *Note*: The first step in the sequence involves a photochemically induced disrotatory electrocyclic ring opening of ergosterol to procalciferol. Vitamin D_2 deficiency leads to rickets, where long bones of the legs bend outwards. This was rife in the UK during the industrial revolution due to poor nutrition and a lack of sunlight in polluted towns.

[3,3]-Sigmatropic rearrangements

These rearrangements involve 6π electrons and suprafacial arrangements for both components in the transition state. Hexa-1,5-diene rearrangements are referred to as *Cope rearrangements*. A variant known as the *oxy-Cope rearrangement* results in the formation of γ,δ-unsaturated carbonyl compounds.

Rearrangement of vinyl allyl ethers is referred to as the *Claisen rearrangement* and gives rise to γ,δ-unsaturated carbonyl compounds. Phenyl allyl ethers undergo the aromatic variant with the initial product tautomerizing to the *ortho*-allylphenol.

Cheletropic reactions

Cheletropic reactions are pericyclic processes in which concerted formation or cleavage of two σ-bonds attached to the same atom occurs. Examples are the suprafacial expulsion of sulfur dioxide from 2,5-dihydrothiophene-1,1-dioxide and the corresponding antarafacial process with 2,7-dihydrothiepin-1,1-dioxide.

Wagner–Meerwein rearrangements

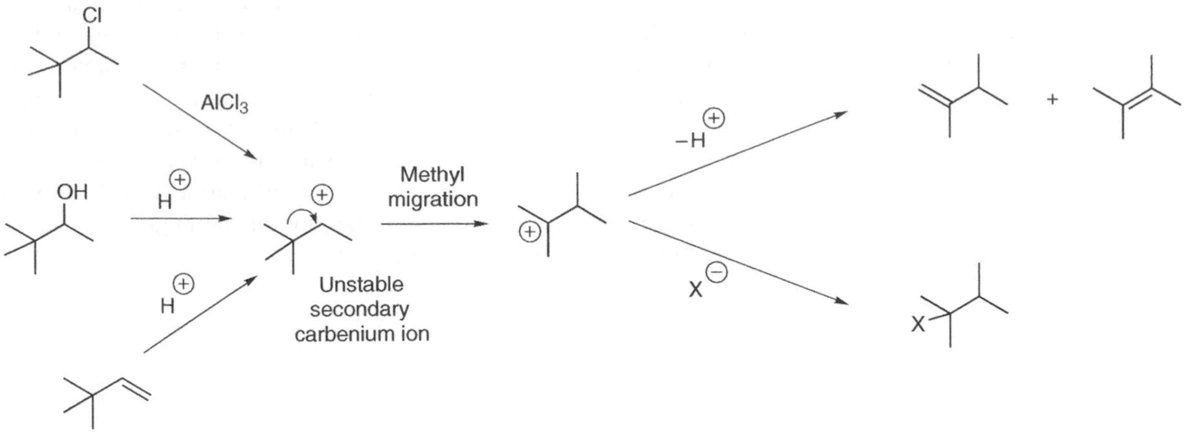

Pinacol rearrangement

Pinacol   *Pinacolone*

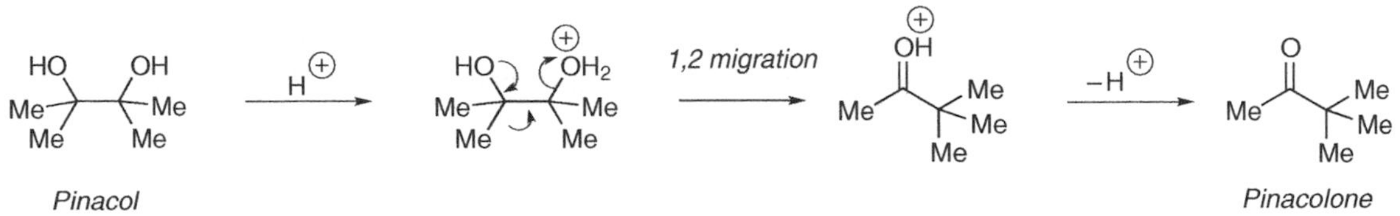

Tiffaneau–Demyanov rearrangement

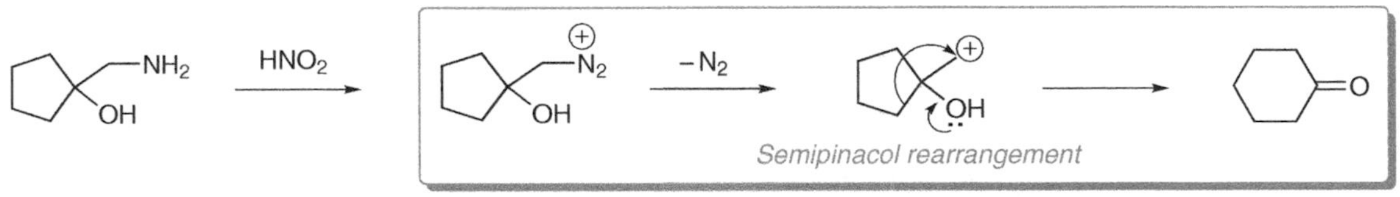

Semipinacol rearrangement

Favorskii rearrangement

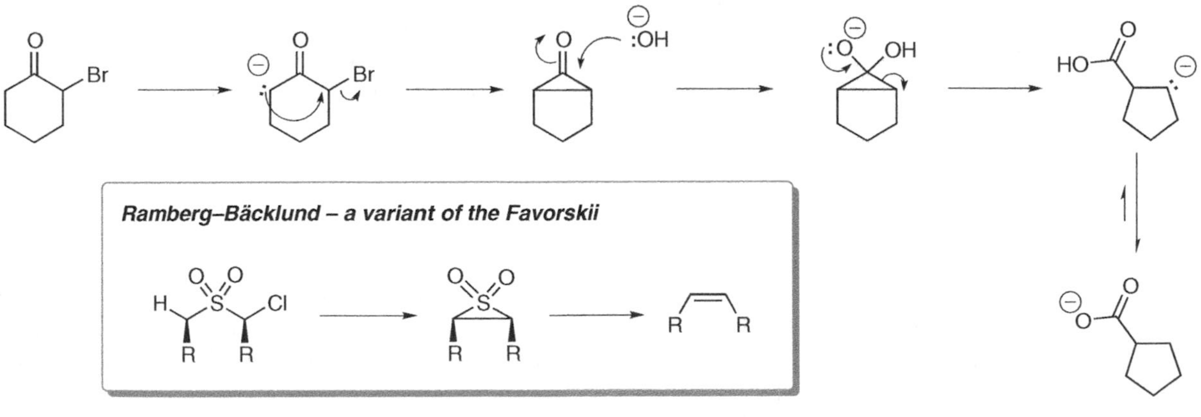

Ramberg–Bäcklund – a variant of the Favorskii

Arndt–Eistert homologation

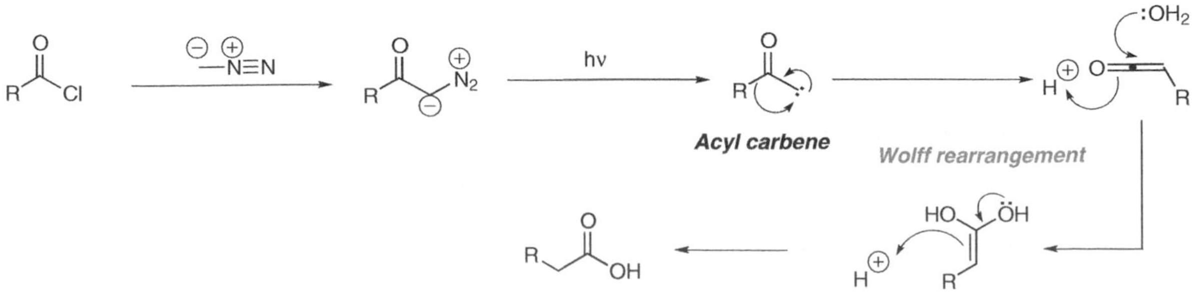

Acyl carbene

Wolff rearrangement

REARRANGEMENTS

5.56 Rearrangements to electron-deficient carbon

The major driving force for 1,2-alkyl migration to a carbocationic centre is formation of a more stable carbocation, sometimes associated with relief from steric compression. With non-racemic substrates intermediacy of a free carbocation at either the *migration terminus* or *migration origin* results in loss of stereochemical integrity at that centre; whereas a concerted migration results in inversion of the absolute stereochemistries of the migration origin and terminus. A substituent capable of stabilizing an adjacent carbocation by acting as an intramolecular nucleophile is said to undergo *neighbouring group participation* or *anchimeric assistance*. Any β-substituent possessing an unshared electron pair or a double bond is a candidate for such participation. The ease with which any group will undergo an 1,2-shift (*migratory aptitude*) may be determined by analyzing the product mixtures obtained from a substrate containing two different competing substituents at the migrating origin.

Wagner–Meerwein and related rearrangements

In 1889 Georg Wagner recognized that problems associated with structural determination of terpenes were due to skeletal rearrangements and in 1922 Hans Meerwein proposed that these were initiated by cationic species. In the archetypal *Wagner–Meerwein rearrangement* the initial carbenium ion is commonly produced by protonation of an alcohol and subsequent loss of water, although halide loss or protonation of a double bond may also promote rearrangement to provide a more stable carbenium ion. This may be quenched by addition of a nucleophile or may undergo deprotonation, forming the most substituted alkene if there is a choice of product. Further rearrangement of the product carbenium may occur in more complex systems and such cascades of rearrangements are responsible for reaction observed in the terpene series. Although fundamentally the same process, 1,2-methyl shifts are known as the *Nametkin rearrangement*.

The carbenium ion produced by 1,2-alkyl shift on treatment of an 1,2-diol with acid undergoes deprotonation of the hydroxy group to furnish a carbonyl group which acts as the driving force for the reaction. This process is known as the *pinacol rearrangement* (the trivial term for a symmetrical 1,2-diol is a *pinacol*); whereas in the *semipinacol rearrangement*, the initial carbenium ion may be generated by loss of halide, protonation of a double bond, acid catalyzed epoxide ring opening or expulsion of nitrogen from a diazonium salt. A variant of the latter process, resulting in ring expansion, is the *Tiffaneau–Demyanov rearrangement*, used to homologate cyclic ketones.

5.57 Base promoted nucleophilic rearrangements

The *Favorskii rearrangement*, named after its discoverer who showed in 1894 that dichloroketones could be converted into unsaturated carboxylic acids on treatment with a base, encompasses base promoted rearrangements of α-chloroketones and is of particular use for ring contraction of cyclic substrates. Enolate generation results in intramolecular displacement of halide to generate a cyclopropanone intermediate which undergoes nucleophilic ring opening to generate the carboxylic acid derivative. Evidence for passage via a symmetrical cyclopropanone intermediate is provided by labelling studies.

An analogous reaction which proceeds by formation of a cyclic *episulfone* intermediate is the base promoted *Ramberg–Bäcklund rearrangement* of α-halo sulfones. However, in this case, cheletropic extrusion of SO_2 (see Section 5.55) results in alkene formation.

5.58 Carbene promoted nucleophilic rearrangements

α-Diazoketones are more stable than their simple aliphatic counterparts and may be isolated. However, heating results in *Wolff rearrangement* by expulsion of nitrogen to generate an acyl carbene (see Section 5.33) which undergoes spontaneous rearrangement to a ketene. Subsequent nucleophilic attack by the solvent forms either a carboxylic acid or ester. The process is particularly useful for ring contraction of cyclic α-diazoketones.

The *Arndt–Eistert homologation* refers to a procedure in which a carboxylic acid is converted via its acid chloride to a diazoketone on treatment with diazomethane. Such substrates undergo Wolff rearrangement in the presence of water to produce carboxylic acids containing an additional carbon compared to the starting acid, permitting stepwise ascent of the homologous series.

Beckmann rearrangement

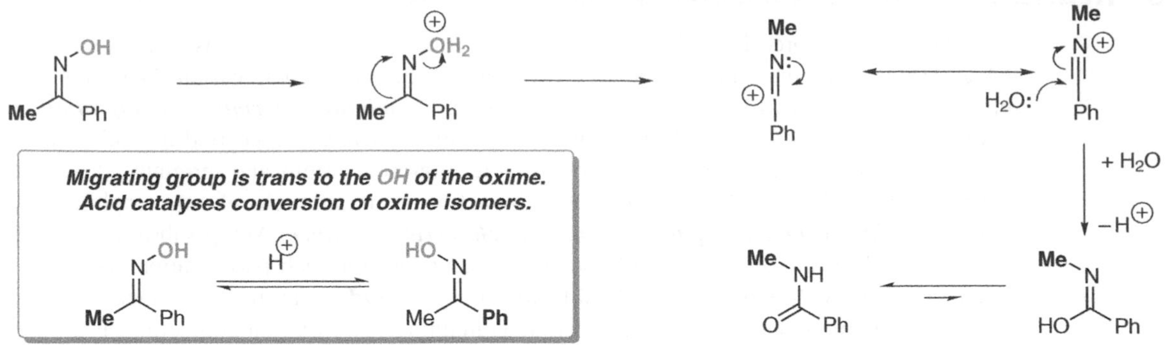

Migrating group is trans to the OH of the oxime.
Acid catalyses conversion of oxime isomers.

Curtius, Lössen, Schmidt rearrangements

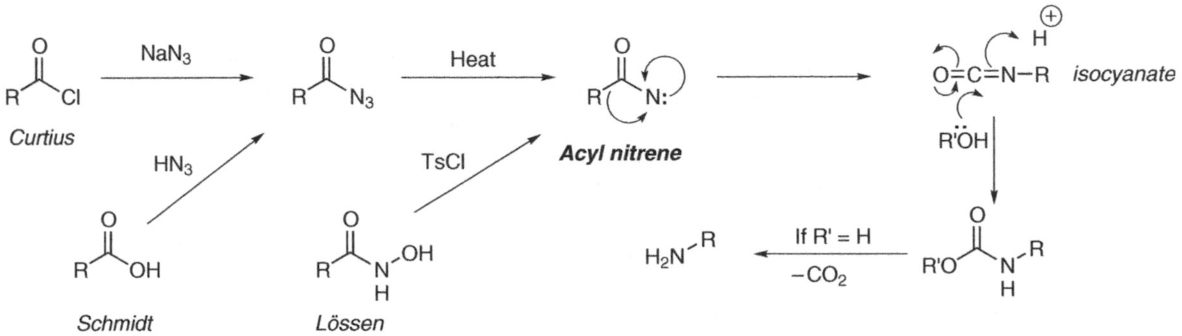

Curtius

Acyl nitrene

isocyanate

Schmidt *Lössen*

Hofmann degradation

isocyanate

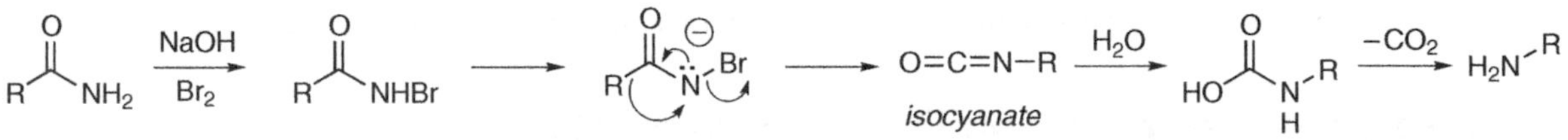

Baeyer–Villiger rearrangement

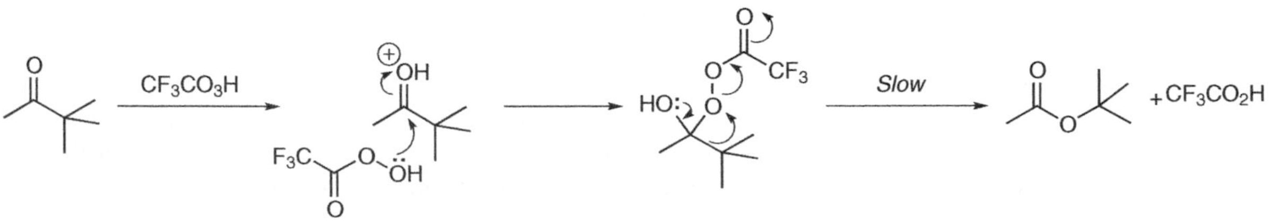

Migratory aptitude: 3°-alkyl > 2°-alky ≈ Ph >1°-alkyl > Me

Hydroperoxide degradation

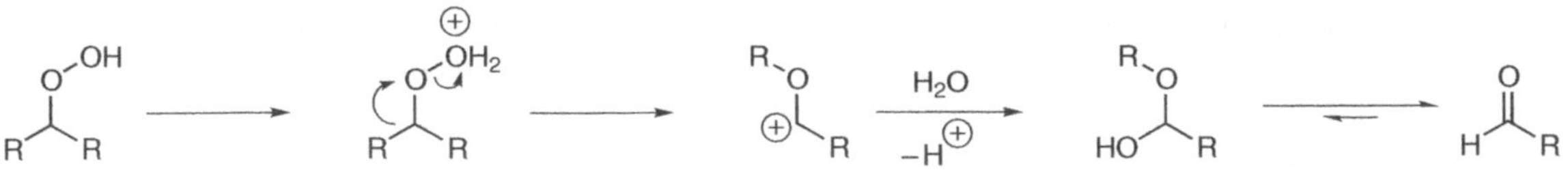

5.59 Rearrangements to electron-deficient nitrogen

Beckmann rearrangement

Under conditions which convert the hydroxyl group into a leaving group, oximes undergo concerted stereospecific migration of the alkyl group *anti-* to the hydroxyl group. The resultant *N*-acyl nitrilium species is attacked by water to give a secondary amide. Although the reaction is stereospecific, due to the configurational lability of oximes under acidic conditions, any stereochemical information is frequently masked by isomerization prior to migration.

Curtius, Lössen, Schmidt rearrangements

This group of very closely related reactions involve 1,2-alkyl shifts induced by the generation of a *nitrene* via decomposition of an azide. Rearrangements which pass via an acyl azide only differ in the carboxylic acid derivative used as the ultimate precursor. In all cases rearrangement occurs in concert with loss of nitrogen from the azide to generate an isocyanate. This is trapped by the solvent to form either a carbamate ester, if an alcohol is used, or the corresponding carbamic acid if carried out under aqueous conditions. Carbamic acids subsequently fragment to liberate carbon dioxide and a primary amine.

The Schmidt rearrangement is more general and carbonyl compounds, alcohols and alkenes can also be used as substrates. With ketones the product is a secondary amide providing an alternative to the Beckmann rearrangement. Rearrangement of tertiary alkyl azides may proceed via an intermediate *nitrene* (see Section 5.34).

Hofmann degradation

Treatment of primary amides with bromine and aqueous sodium hydroxide (a source of NaOBr) produces *N*-bromoamides which subsequently undergo deprotonation, prompting carbon-to-nitrogen alkyl migration with loss of bromide. The resultant isocyanate is then trapped by water and the carbamic acid fragments to give a primary amine. The migration is probably concerted but stepwise loss of bromide to generate an intermediate *nitrene* cannot be ruled out.

5.60 Rearrangements to electron-deficient oxygen

This small group of reactions involves heterolytic cleavage of a peroxide O–O bond in concert with an alkyl group migration and parallels carbon-to-nitrogen rearrangements. The leaving group is commonly a carboxylic acid or its conjugate base. Concerted alkyl group migration avoids the formation of an unstable positively charged oxygen species. Such reactivity is at variance with the standard reactivity of the peroxide linkage which usually undergoes homolytic cleavage.

Baeyer–Villiger rearrangement

This is the most general and synthetically important carbon-to-oxygen rearrangement. Nucleophilic addition of a peracid to a ketone forms a tetrahedral *Criegee intermediate* which collapses with rate-determining expulsion of the carboxylic acid and concerted migration of one of the alkyl groups to generate an ester. Cyclic ketones form lactones. An important feature of the rearrangement is that unsymmetrical ketones show highly selective migration of the group which is more highly substituted at the α-position to the carbonyl group. As loss of carboxylic acid or carboxylate occurs in the rate-determining step, the overall reaction rate is dictated by the leaving group ability. Thus, trifluoroperacetic acid is usually the most effective reagent.

An analogous reaction in which aminoaryl or hydroxyaryl ketones rearrange on treatment with alkaline hydrogen peroxide is known as the *Dakin reaction*.

Hydroperoxide degradation

Protonation of a hydroperoxide results in loss of water and migration of an alkyl group to oxygen, generating a carbenium ion which is quenched by water to form a *hemiacetal* (see Section 5.41) which is hydrolyzed to a ketone under the reaction conditions.

6

Compound Classes

6.1 Functional group chemistry

Alkane synthesis

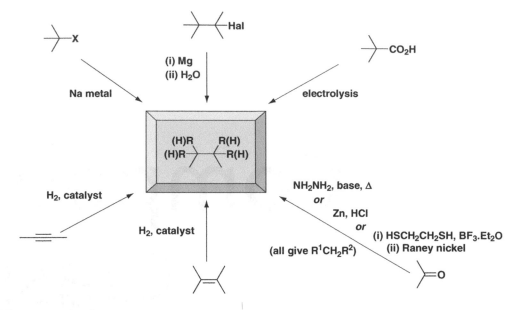

Alkene synthesis

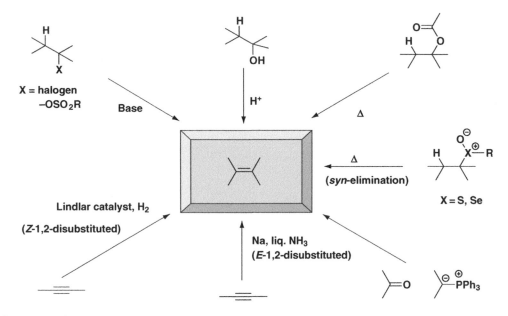

Alkyne synthesis

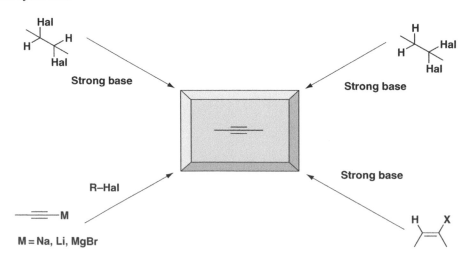

Halide synthesis

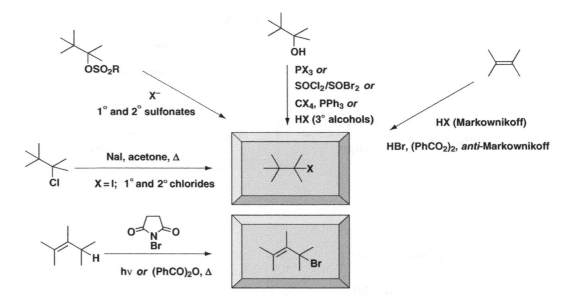

Alcohol synthesis

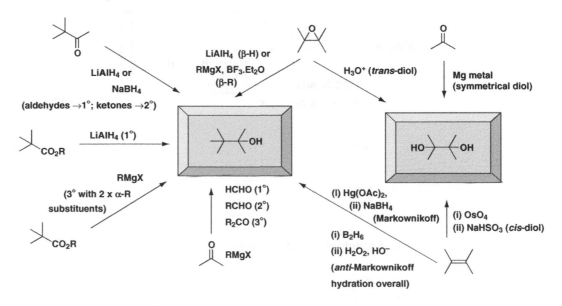

Ether synthesis

Aldehyde synthesis

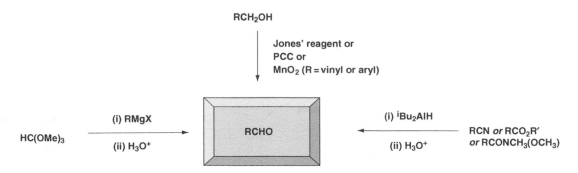

Ketone synthesis

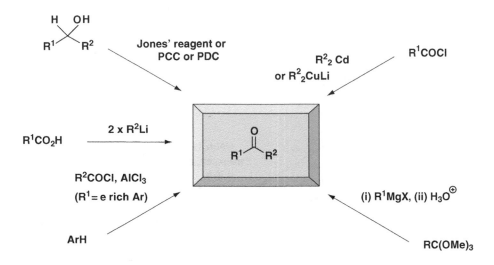

Reactions of aldehydes and ketones

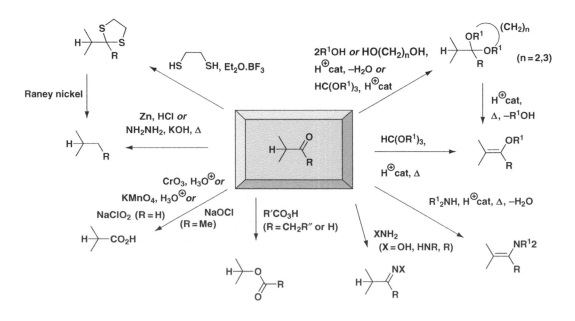

Carboxylic acid synthesis

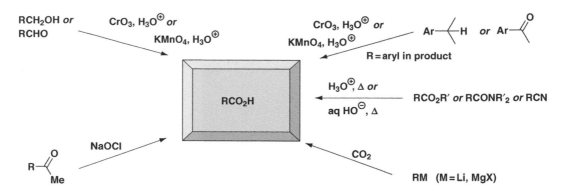

Reactions of carboxylic acids

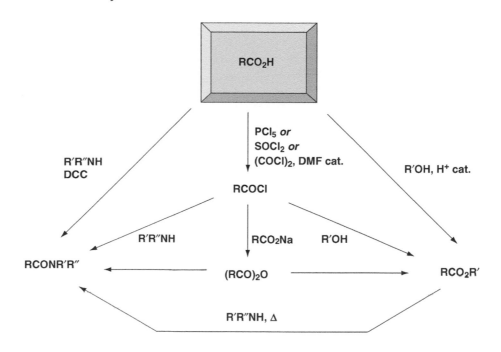

Amine synthesis

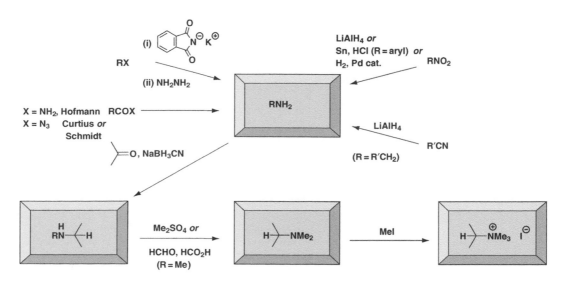

Index

Index

Printed and bound in the UK by
CPI Antony Rowe, Eastbourne

FRONT BOOKBINDING OF THIS PUBLICATION

DATE DUE

Printed and bound by CPI Group (UK) Ltd, Croydon, CR0 4YY

27/10/2024

14580201-0005